Reliable Paradigm®

AF812683

High Performance Database Tuning & Optimization

Unlock Database Performance Mastery: A Comprehensive Guide

Includes Optimization for Cloud Computing, AI, Machine-learning, Predictive Analytics, and Large Language Models

Marek Kos

High Performance Database Tuning & Optimization

by Marek Kos

Published by Reliable Paradigm LLC

Reliable Paradigm books may be purchased for educational, business, or sales promotional use. Online editions are also available for most titles. For more information, contact our corporate / institutional sales department at our website: ReliableParadigm.com

Revision History:

2023-November: Initial Revision

Web Page and Errata: reliableparadigm.com/high-performance-database-tuning-optimization

ISBN: **978-1-962939-00-3**

TABLE OF CONTENTS

CHAPTER 1. INTRODUCTION

THE IMPORTANCE OF DATABASE PERFORMANCE AND ITS OPTIMIZATION

Databases serve as the backbone of modern applications, storing and managing vast amounts of valuable data. However, the mere existence of a database is insufficient; its performance plays a crucial role in ensuring the smooth operation of applications and the satisfaction of end-users. Database performance optimization is the process of enhancing efficiency, responsiveness, and scalability, enabling businesses to maximize their data-driven capabilities. This essay explores the significance of database performance and why investing in its optimization is a fundamental aspect of any successful organization. Database performance optimization is not a luxury; it is a necessity in today's data-driven world. The importance of efficient data retrieval, scalability, cost-efficiency, informed decision-making, reliability, and data integrity cannot be overstated. Investing in database performance optimization enables businesses to deliver seamless user experiences, support growth and scalability, reduce costs, and unlock the full potential of their data assets. As organizations continue to navigate the ever-expanding realm of data, prioritizing database performance optimization will remain an essential element in achieving operational excellence and driving success.

Understanding the key factors impacting database performance is crucial for building and maintaining highly efficient database systems. From proper database design and schema considerations to effective query optimization, storage and disk I/O optimization, memory and caching strategies, and system configuration tuning, each factor plays a significant role in enhancing database performance. By addressing these factors and implementing appropriate optimization techniques, organizations can achieve faster query response times, improved scalability, and a seamless user experience, ultimately unlocking the full potential of their data assets.

This text is a practical guide that explores the fundamentals, strategies, and advanced techniques for maximizing database performance, covering topics such as database design optimization, query tuning, hardware considerations, application-database interaction, scalability, performance testing, and emerging trends in the field, providing readers with a holistic understanding of database performance optimization.

EFFICIENT DATA RETRIEVAL AND PROCESSING

In today's fast-paced digital landscape, users expect instantaneous access to information. A well-optimized database ensures quick data retrieval, enabling applications to respond promptly to user queries and requests. By minimizing response times, businesses can enhance user experience, retain customers, and gain a competitive edge. Whether it's an e-commerce platform, a banking application, or a customer relationship management system, optimized database performance is vital for delivering real-time data access and processing.

SCALABILITY AND GROWTH

As organizations expand, their data volumes increase exponentially. A robust database performance optimization strategy enables seamless scalability, accommodating growing data needs without compromising system efficiency. By efficiently managing storage, memory, and processing resources, businesses can support the influx of data and users, ensuring that performance remains consistent even under heavy workloads. Scalability is crucial for handling spikes in demand, maintaining user satisfaction, and supporting business growth.

COST-EFFICIENCY

Database performance optimization can significantly impact cost-efficiency. A well-tuned database maximizes hardware utilization and minimizes resource waste, reducing the need for additional infrastructure investments. By optimizing query execution plans, indexing strategies, and caching mechanisms, organizations can enhance application performance while minimizing hardware and operational expenses. Improved performance also translates to reduced downtime, minimizing revenue loss and potential business disruptions.

DECISION-MAKING AND ANALYTICS

Data-driven decision-making is a cornerstone of successful organizations. A high-performance database enables timely and accurate data processing, facilitating real-time analytics and insights. When databases are optimized, the extraction, transformation, and loading (ETL) processes can be expedited, enabling faster access to updated data for analysis. By empowering decision-makers with up-to-date and reliable information, businesses can make informed strategic choices, identify trends, detect anomalies, and gain a competitive advantage.

RELIABILITY AND DATA INTEGRITY

A well-optimized database not only ensures speed and efficiency but also upholds data integrity and reliability. Database performance optimization includes implementing robust data validation and integrity checks, enforcing proper constraints, and ensuring data consistency. By maintaining data integrity, businesses can rely on their databases as trusted sources of information, promoting trust with customers, partners, and regulatory bodies.

KEY FACTORS IMPACTING DATABASE PERFORMANCE

Database performance is a critical aspect of any modern information system, as it directly affects the efficiency, responsiveness, and reliability of data storage and retrieval. To optimize database performance effectively, it is essential to understand the key factors that influence its overall speed and functionality. This essay explores the primary factors that impact database performance and highlights their significance in maintaining robust and efficient database systems.

1. DATABASE DESIGN AND SCHEMA

The design and schema of a database play a pivotal role in its performance. A database schema is a logical representation of the structure and organization of a database. It defines the layout, relationships, and constraints of the data stored in the database. In other words, a schema acts as a blueprint or plan that outlines how data will be organized, stored, and accessed within the database.

The database schema includes the following key elements:

1. **Tables**: are the main building blocks of a database schema. Each table represents a specific entity or concept and consists of rows (records) and columns (attributes). For example, a database for an online store may have tables for customers, products, orders, and shipments.

2. **Columns** (Attributes): define the individual data elements that make up a table. Each column corresponds to a specific piece of information or characteristic associated with the entity represented by the table. For instance, a "Customers" table may have columns such as "CustomerID," "Name," "Email," and "Address."

3. **Relationships**: establish connections between tables and define how the data in different tables are related to one another. Relationships can be one-to-one, one-to-many, or many-to-many, and they are typically defined through foreign keys that link data between tables.

4. **Constraints**: are rules that enforce data integrity and govern the allowed values or behavior within the database. Common constraints include primary keys, which ensure uniqueness of records, and foreign keys, which maintain referential integrity between related tables.

5. **Views and Procedures**: In addition to tables, a database schema may also include views and procedures. Views are virtual tables that provide a specific subset of data from one or more tables, while procedures are predefined sets of SQL statements that can be executed as a single unit to perform specific tasks.

6. **Indexes**: are optional database schema elements that improve data retrieval performance by providing quick access to specific data. They create a data

structure that allows the database management system to locate information more efficiently.

Relational modeling, also known as entity-relationship modeling, is a conceptual data modeling technique used to design and represent the structure and relationships of data in a relational database. It is a fundamental process in the design of relational databases, which are based on the relational model, a theoretical framework introduced by Edgar F. Codd in the 1970s.

Relational modeling involves identifying and defining entities (also known as tables) and their attributes (also known as columns) and establishing relationships between entities. The process typically involves the following key concepts:

1. **Entities**: represent real-world objects or concepts that need to be stored and managed in the database. Each entity corresponds to a table in the database and contains attributes that describe the characteristics of the entity. For example, in a university database, entities could include "Student," "Course," and "Instructor."

2. **Attributes**: represent the specific properties or characteristics of an entity. Each attribute corresponds to a column in the table representing the entity. For instance, attributes for the "Student" entity could include "Student ID," "Name," "Age," and "Major."

3. **Relationships**: define how entities are related to one another. They establish associations between different entities in the database. Relationships can be one-to-one, one-to-many, or many-to-many, indicating the cardinality between entities. For example, a student can enroll in multiple courses (one-to-many), and a course can have multiple enrolled students (also one-to-many).

4. **Primary Keys**: are unique identifiers that uniquely identify each record (row) in an entity. They ensure that each entity instance is distinct and can be referenced from other entities through foreign keys. Primary keys play a crucial role in maintaining data integrity and establishing relationships between tables.

5. **Foreign Keys**: are attributes in one entity that refer to the primary key of another entity. They facilitate the establishment of relationships between entities and enable data integrity constraints, such as referential integrity, to maintain consistency in the database.

Well-designed databases with normalized schemas promote data integrity, reduce redundancy, and facilitate efficient querying. Properly defined relationships, indexes, and constraints enable the database to process queries more rapidly, resulting in improved performance.

2. INDEXING AND QUERY OPTIMIZATION

The efficiency of queries greatly impacts database performance. A database query is a request or command that is sent to a database management system (DBMS) to retrieve, manipulate, or modify data stored in a database. Queries are written in a specific database query language, such as Structured Query Language (SQL), which is a standard language used for interacting with relational databases.

A database query allows users to:

1. **Retrieve Data**: Queries can be used to retrieve specific data from one or more tables in the database based on specified criteria. For example, a simple SELECT query in SQL can retrieve all the records from a "Customers" table where the "City" is "New York."

2. **Filter Data**: Queries can filter data based on specific conditions or criteria. For instance, a SELECT query with a WHERE clause can filter records that meet certain conditions, such as all "Completed" orders from a "Sales" table.

3. **Perform Aggregations**: Queries can perform aggregate functions on data, such as calculating the sum, average, maximum, or minimum values of a particular column. For example, a query can calculate the total sales revenue for a specific period.

4. **Join Data**: Queries can join data from multiple tables to create a single result set that combines information from related tables. Joins are used to retrieve data from multiple tables based on common columns.

5. **Update Data**: Queries can be used to modify existing data in the database. For example, an UPDATE query can change the "Status" of an order from "Pending" to "Shipped."

6. **Insert Data**: Queries can add new data to the database. An INSERT query adds new records to a table.

7. **Delete Data**: Queries can remove data from the database. A DELETE query removes specific records from a table.

Database queries are essential for interacting with and managing data in a relational database. They provide a powerful and flexible way to access and manipulate data, making databases valuable tools for storing, organizing, and retrieving information in various applications and industries. The ability to write efficient and effective queries is crucial for ensuring optimal database performance and meeting specific data processing needs.

The use of appropriate indexes on frequently accessed columns significantly speeds up data retrieval, eliminating the need for full table scans. A database index is a data structure that enhances the efficiency and speed of data retrieval operations in a database management system (DBMS). It acts as a lookup table that helps the DBMS quickly locate specific data within a database table, reducing the need for full table scans and improving query performance.

An index is created on one or more columns of a database table, and it contains a sorted list of values from those columns along with pointers to the corresponding rows in the table. When a query is executed that involves the indexed columns, the DBMS can use the index to quickly locate the desired data, reducing the number of disk reads and resulting in faster query execution.

The main characteristics of a database index are:

1. **Improved Query Performance**: By providing direct access to specific data, indexes speed up data retrieval operations. Queries that use indexed columns in their WHERE clauses or involve join operations benefit the most from indexes.

2. **Supported Data Types**: Indexes can be created on various data types, including integers, strings, dates, and more. The choice of columns for indexing depends on the types of queries frequently executed against the table.

3. **Unique and Non-Unique Indexes**: Indexes can be unique or non-unique. A unique index ensures that the indexed column contains only unique values, preventing duplicate entries in the table. On the other hand, non-unique indexes allow duplicate values in the indexed column.

4. **Index Maintenance**: Indexes need to be updated whenever data in the indexed columns is inserted, updated, or deleted. While indexes enhance query performance, they require additional resources and maintenance overhead during data modifications.

5. **Storage Space**: Indexes occupy additional storage space in the database as they create an additional data structure. As a trade-off between query performance and storage, careful consideration is needed when deciding which columns to index.

6. **Composite Indexes**: A composite index is an index created on multiple columns, allowing queries that involve multiple columns to be efficiently optimized.

Database indexes are essential tools for improving the performance of data retrieval operations, especially in large and complex databases. However, their usage should be carefully planned to avoid unnecessary index creation, as they come with additional overhead during data modification operations. Indexing is a fundamental aspect of database performance tuning, and choosing the appropriate columns to index can significantly impact the efficiency of the database management system.

Query optimization techniques, such as rewriting queries, utilizing query execution plans, and employing appropriate join strategies, enhance the performance of complex queries and minimize unnecessary processing. These techniques are examined n detail in a future chapter.

3. STORAGE AND DISK I/O

The underlying storage infrastructure and disk I/O (Input / Output) operations heavily influence database performance. The choice of storage technology, such as solid-state drives (SSDs) or hard disk drives (HDDs), impacts data access speeds. Optimizing disk I/O through techniques like data partitioning and sharding, file placement, and intelligent caching can significantly enhance database performance by minimizing latency and improving data transfer rates.

4. MEMORY AND CACHING

Database systems rely on memory for efficient data retrieval and processing. Utilizing an appropriate amount of memory for caching frequently accessed data minimizes disk I/O operations, leading to faster response times. Techniques like buffer caches, query result caching, and connection pooling optimize memory usage and improve overall performance.

5. CPU UTILIZATION AND PARALLEL PROCESSING

The central processing unit (CPU) is a critical resource for database operations. Utilizing parallel processing techniques, such as parallel query execution, parallelizing data loading, and using multiple CPU cores, enhances performance by distributing the workload and utilizing available resources efficiently.

6. NETWORK AND LATENCY

In distributed database systems or client-server architectures, network performance and latency affect overall database performance. Optimizing network protocols, minimizing data transmission over the network, and reducing network round-trips enhance application responsiveness and user experience.

7. SYSTEM CONFIGURATION AND PARAMETER TUNING

Database systems offer various configuration parameters that can be adjusted to optimize performance. Proper tuning of parameters, such as memory allocation, concurrency settings, buffer sizes, and cache configurations, ensures that the database system is optimized for the specific workload and hardware environment.

CHAPTER 2. FUNDAMENTALS CONCEPTS OF DATABASE PERFORMANCE OPTIMIZATION

To establish a solid foundation in optimizing database performance, it is essential to delve into the fundamentals that underpin efficient data management. This section of the book focuses on the core concepts and principles of performance tuning, highlighting essential areas such as relational modeling, query processing and optimization, memory and disk management, as well as indexing and partitioning strategies. Proceeding chapter sections delve into these topics in greater detail. By mastering these fundamentals, readers will gain a comprehensive understanding of the key components influencing database performance and acquire the necessary skills to optimize their databases effectively.

1. RELATIONAL MODELING:

Relational modeling lays the groundwork for efficient database performance. This topic explores the principles of proper schema design, normalization, and denormalization techniques. By understanding the relational model, readers can create optimized database structures that minimize redundancy, ensure data integrity, and facilitate efficient data retrieval.

Relational modeling provides a clear and structured representation of data and its relationships, making it easier for database administrators and developers to design, implement, and maintain relational databases. The resulting entity-relationship diagrams (ER diagrams) serve as visual representations of the relational model, helping stakeholders understand the database structure and facilitating effective communication during the database design process.

Database normalization is a process in database design that organizes and structures a relational database to minimize data redundancy and dependency issues. The primary goal of normalization is to eliminate data anomalies and improve data integrity, consistency, and efficiency within the database.

The normalization process involves breaking down a large table into smaller, more manageable tables, each with a specific purpose and relationship to one another. This is achieved through a series of normalization rules, commonly referred to as normal forms, which define the criteria for organizing the data.

There are different levels of normalization, known as normal forms, which progressively reduce data redundancy and dependency:

a. First Normal Form (1NF): This is the most basic level of normalization. It ensures that each column in a table holds atomic values, meaning each cell contains a single value, and there are no repeating groups of data. By adhering to 1NF, each entry in the table becomes unique, facilitating efficient data retrieval and manipulation.

b. Second Normal Form (2NF): To achieve 2NF, a table must first be in 1NF. Additionally, all non-key attributes (attributes that are not part of the primary key) must be fully functionally dependent on the entire primary key. This eliminates partial dependencies and reduces data redundancy by ensuring that attributes depend only on the primary key.

c. Third Normal Form (3NF): To reach 3NF, a table must be in 2NF and must eliminate transitive dependencies. Transitive dependencies occur when an attribute depends on another non-key attribute rather than directly on the primary key. Breaking down tables to eliminate these dependencies improves data integrity and reduces redundancy.

There are higher levels of normalization beyond 3NF, such as Boyce-Codd Normal Form (BCNF) and Fourth Normal Form (4NF), which further refine the database structure to address more complex relationships and dependencies.

Database normalization is a crucial step in the database design process as it ensures data consistency, reduces data duplication, and simplifies data management. By adhering to normalization rules, database designers create a more robust and maintainable database schema that facilitates efficient data retrieval and modification while minimizing the risk of data anomalies.

Database denormalization is the process of intentionally introducing redundancy into a relational database by reversing the normalization process. Unlike normalization, which aims to eliminate data redundancy and ensure data integrity, denormalization deliberately duplicates data to improve query performance and simplify certain types of data retrieval operations.

Denormalization is often employed in situations where read performance is prioritized over data modification performance and storage space concerns. By storing redundant data in the database, queries that require extensive joins and data retrieval from multiple tables can be simplified, leading to faster query execution.

There are several denormalization techniques that can be applied to a database:

a. Flattening: Flattening involves combining data from multiple related tables into a single table. This eliminates the need for complex joins during query execution, which can significantly improve query performance.

b. Materialized Views: Materialized views are precomputed, stored subsets of data resulting from complex queries. These views are updated periodically, reducing the need to compute the same aggregations or calculations repeatedly during query execution.

c. Redundant Columns: Redundant columns involve duplicating data from one table into another related table. This technique can be useful when data from multiple tables is frequently queried together, reducing the number of joins needed for these queries.

d. Partial Denormalization: Partial denormalization involves denormalizing only specific portions of a database while maintaining normalization for other parts. This allows for a balance between performance and data integrity in different areas of the database.

While denormalization can improve query performance, it also comes with some trade-offs and potential challenges:

a. Data Redundancy: Introducing redundant data can lead to an increased risk of data inconsistency if updates are not properly managed. Maintaining data integrity becomes the responsibility of the application logic, rather than relying solely on the database schema.

b. Increased Storage: Denormalization can lead to increased storage requirements since redundant data is stored multiple times. This may be a concern for large databases with limited storage resources.

c. Complexity: Denormalized databases can become more complex to manage and maintain, especially when updates involve changes to redundant data in multiple locations.

d. Maintenance: Updates to denormalized databases require careful attention to ensure data consistency across all redundant copies.

Database denormalization should be approached with caution and used judiciously, as it introduces a trade-off between improved query performance and increased complexity and maintenance overhead. It is essential to carefully assess the specific requirements of the database and the nature of the queries to determine the appropriate level of denormalization to be applied.

2. QUERY PROCESSING AND OPTIMIZATION:

Query processing and optimization techniques are essential for enhancing database performance. This section covers the inner workings of query execution plans, the role of query optimizers, and strategies for optimizing queries. Readers will learn how to interpret query execution plans, identify performance bottlenecks, and employ optimization techniques such as index selection, join optimization, and query rewriting to improve query performance.

3. MEMORY AND DISK MANAGEMENT:

Efficient memory and disk management are vital for database performance. This topic explores strategies for managing memory resources effectively, including buffer caches, data caching, and optimal memory allocation. Additionally, readers will learn about disk management techniques, such as read-ahead caching, data placement, and I/O optimization, to minimize disk latency and maximize data retrieval speed.

4. INDEXING AND PARTITIONING STRATEGIES:

Indexes and partitions are essential components for optimizing data access and storage in databases. This section delves into various indexing techniques, including B-trees, hash indexes, and bitmap indexes, explaining when and how to use them effectively. Partitioning strategies, such as range partitioning, list partitioning, and hash partitioning, are also covered, providing insights into data distribution and query performance improvements.

In addition to the fundamentals mentioned earlier, several other performance tuning concepts and principles are relevant for optimizing database performance. These include the following topics that are addressed in greater detail in proceeding chapters:

A. STATISTICS AND DATA DISTRIBUTION:

Accurate and up-to-date statistics about data distribution and cardinality are crucial for the database optimizer to make informed decisions. Database cardinality refers to the number of distinct values or unique entries in a column of a database table. It represents the uniqueness or diversity of data in a specific column and helps in understanding the data distribution. Cardinality is an important concept in database management systems and is used by query optimizers to make efficient decisions while processing queries.

- Low Cardinality: A column with low cardinality contains only a few distinct values relative to the total number of rows in the table. For example, a column representing "Gender" in a dataset containing millions of records may have only two distinct values, "Male" and "Female." Low cardinality columns are typically not useful for indexing as they do not provide significant selectivity.

- Medium Cardinality: A column with medium cardinality has a moderate number of distinct values relative to the total number of rows. An example of a medium cardinality column could be a "Country" column in a large dataset containing thousands of distinct country names. Medium cardinality columns may provide some selectivity and can be used for indexing in certain scenarios.

- High Cardinality: A column with high cardinality contains a large number of distinct values compared to the total number of rows. For instance, a column representing "Email addresses" in a database with millions of records would likely have a vast number of unique values. High cardinality columns are generally suitable for indexing, as they offer significant selectivity and help improve query performance. Analyzing and maintaining statistics on tables, indexes, and columns enable the optimizer to generate optimal query execution plans. Techniques like histogram creation, data sampling, and automatic statistics updates help ensure accurate statistics for efficient query processing.

B. QUERY REWRITING AND OPTIMIZATION:

Advanced query rewriting techniques can significantly impact performance. This involves transforming queries into alternative forms that are more efficient to execute. Common techniques include query simplification, subquery optimization, predicate (query condition) pushdown, and query transformations based on semantic equivalences. By rewriting and optimizing queries, database systems can achieve faster and more efficient data retrieval.

C. CONCURRENCY CONTROL AND LOCKING:

Concurrency control mechanisms ensure data consistency in multi-user environments. Optimizing concurrency control involves selecting appropriate isolation levels, managing locks efficiently, and utilizing lock-free data structures where applicable. Techniques such as lock escalation, lock granularities, and concurrency control algorithms like multi-version concurrency control (MVCC) are essential for minimizing contention and maximizing parallelism.

D. DATABASE CONFIGURATION AND PARAMETER TUNING:

Database systems offer a multitude of configuration parameters that can be tuned to optimize performance. Understanding these parameters and their impact on system behavior is crucial. Examples include buffer pool size, cache settings, thread concurrency, and network-related parameters. Fine-tuning these settings based on workload characteristics and hardware capabilities can significantly enhance overall performance.

E. QUERY PERFORMANCE MONITORING AND ANALYSIS:

Monitoring and analyzing query performance is essential for identifying performance bottlenecks and areas for improvement. Techniques such as query profiling, tracing, and capturing execution plans help pinpoint problematic queries. By analyzing query performance metrics, execution times, resource usage, and I/O patterns, database administrators can identify and optimize queries that are consuming excessive resources.

F. DATA COMPRESSION AND STORAGE OPTIMIZATION:

Data compression techniques reduce storage requirements and enhance data retrieval speeds. Various compression algorithms, such as gzip, Snappy, or columnar compression, can be applied to tables, indexes, or specific columns. Additionally, optimizing storage structures, such as page size and block sizes, can improve I/O performance and reduce disk space usage.

G. PERFORMANCE TESTING AND BENCHMARKING:

Performing thorough performance testing and benchmarking is crucial for evaluating and fine-tuning database performance. By simulating real-world workloads, measuring response times, and comparing performance metrics, organizations can identify performance bottlenecks, validate optimization efforts, and make data-driven decisions for further enhancements.

H. CONTINUOUS MONITORING AND PERFORMANCE TUNING:

Database performance is not a one-time optimization effort; it requires continuous monitoring and tuning. Implementing monitoring tools, setting up alerts, and establishing performance baselines enable proactive identification and resolution of performance issues. Regular database maintenance tasks, such as index reorganization, statistics updates, and database reorganizations, help sustain optimal performance over time.

By considering these additional performance tuning concepts and principles, database administrators and developers can take a comprehensive approach to optimize database performance, ensuring efficient data processing, faster query response times, and a smooth user experience.

The fundamentals of database performance optimization lay the groundwork for building efficient and high-performing databases. By understanding relational modeling principles, query processing and optimization techniques, memory and disk management strategies, as well as indexing and partitioning approaches, readers can gain a solid foundation for improving database performance. Armed with this knowledge, they will be well-equipped to tackle more advanced topics and implement optimization strategies tailored to their specific database systems, ultimately maximizing the speed, scalability, and responsiveness of their databases.

UNDERSTANDING DATABASE PERFORMANCE METRICS AND BENCHMARKS

Database performance metrics and benchmarks play an instrumental role in assessing and optimizing the efficiency and effectiveness of database systems. These metrics provide valuable insights into various aspects of database performance, allowing organizations to identify bottlenecks, measure system responsiveness, and compare performance against industry standards. This section delves into the significance of understanding database performance metrics and benchmarks, exploring their key components, methodologies, and how they contribute to the optimization of database systems.

1. DATABASE PERFORMANCE METRICS:

Database performance metrics are quantitative measures that gauge the efficiency, responsiveness, and overall health of a database system. These metrics help evaluate fundamental aspects such as query execution times, transaction throughput, resource utilization, and system availability. Some commonly used performance metrics include:

- Response Time: The time taken to complete a database operation, such as query execution or transaction processing. It reflects the system's ability to quickly retrieve and manipulate data.
- Throughput: The rate at which the database processes transactions or queries within a given time frame. It measures the system's processing capacity and scalability.
- CPU Utilization: The percentage of time the CPU spends executing database-related tasks. High CPU utilization may indicate heavy processing or inefficient query execution.
- Memory Usage: The amount of memory utilized by the database for caching, query execution, and other operations. Monitoring memory usage helps optimize caching and allocation strategies.
- Disk I/O: The rate at which data is read from or written to disk. Disk I/O metrics, such as read/write latency and IOPS (Input/Output Operations Per Second), provide insights into storage performance.

2. DATABASE PERFORMANCE BENCHMARKS:

Database performance benchmarks are standardized tests designed to measure the performance of database systems against predefined workloads and scenarios. These benchmarks serve as industry benchmarks for comparing database performance across different platforms and configurations. Commonly used database performance benchmarks include:

- Transaction Processing Performance Council (TPC): TPC benchmarks evaluate database systems' ability to handle transactional workloads and measure performance metrics like throughput, response times, and price-performance ratios.
- Standard Performance Evaluation Corporation (SPEC): SPEC benchmarks focus on performance evaluation of various computing systems, including database servers. They provide standardized workloads and metrics to assess performance across different hardware and software configurations.

- TPC-H and TPC-DS: These benchmarks simulate decision support and analytical workloads, respectively. They evaluate database systems' ability to handle complex queries, data aggregations, and join operations efficiently.
- YCSB (Yahoo! Cloud Serving Benchmark): YCSB focuses on measuring the performance of cloud-based database systems, emphasizing scalability, latency, and read/write operations.

IMPORTANCE OF DATABASE PERFORMANCE METRICS AND BENCHMARKS

Understanding database performance metrics and benchmarks is essential for several reasons:

- **Performance Analysis**: Metrics help identify performance bottlenecks, analyze system behavior, and guide optimization efforts. They provide insights into resource utilization, query efficiency, and system responsiveness.
- **Comparison and Evaluation**: Benchmarks facilitate fair comparisons among different database systems, helping organizations select the most suitable platform for their specific requirements. They enable performance evaluations and inform decisions related to hardware and software upgrades.
- **Capacity Planning**: Performance metrics aid in capacity planning by predicting future resource needs based on historical data. They help organizations allocate appropriate hardware resources and plan for future scalability.
- **Continuous Optimization**: Regularly monitoring and analyzing performance metrics enable ongoing optimization efforts. Identifying trends, spotting anomalies, and setting performance baselines allow for proactive identification and resolution of performance issues.

Understanding database performance metrics and benchmarks is vital for optimizing database systems, ensuring efficient data processing, and delivering a seamless user experience. These metrics provide valuable insights into various aspects of database performance, allowing organizations to identify bottlenecks.

COMMON PERFORMANCE BOTTLENECKS: CAUSES AND PREVENTION STRATEGIES

In the realm of database performance optimization, identifying and addressing performance bottlenecks is requisite for maintaining efficient and responsive systems. Performance bottlenecks are areas within a database system where resources are strained, resulting in degraded performance. Highlighted below are common performance bottlenecks, we explore their underlying causes, and provides practical tips to prevent or mitigate these issues, enabling optimal database performance:

1. **Disk I/O Bottlenecks**:

Causes:

- Inefficient disk access patterns, such as full table scans or excessive random disk reads.
- Insufficient disk bandwidth or slow disk response times.
- Lack of proper disk partitioning and data placement strategies.

Prevention:

- Utilize appropriate indexing strategies to minimize disk reads and optimize data retrieval.
- Implement data partitioning techniques to distribute data across multiple disks or disk arrays.
- Employ caching mechanisms to reduce the frequency of disk I/O operations.
- Monitor disk performance metrics and consider upgrading to faster storage solutions, such as solid-state drives (SSDs).

2. **CPU Bottlenecks**:

Causes:

- Poorly optimized queries or inefficient query execution plans.
- Intensive CPU-bound operations, such as complex calculations or data transformations.
- Insufficient CPU resources or lack of parallel processing capabilities.

Prevention:

- Optimize queries through proper indexing, query rewriting, and query plan analysis.
- Identify and optimize CPU-intensive operations, offloading them to specialized hardware or optimized algorithms.
- Utilize parallel query execution techniques to distribute processing load across multiple CPU cores.

- Consider upgrading to a more powerful CPU or scaling up the hardware infrastructure.

3. **Memory Bottlenecks**:

Causes:

- Insufficient memory allocation for database caching and query processing.
- Inefficient memory management leading to excessive swapping or paging.
- Large result sets or inefficient data retrieval resulting in excessive memory usage.

Prevention:

- Properly configure memory settings, allocating sufficient memory for database caching and query workloads.
- Implement effective caching strategies, utilizing buffer caches and query result caching.
- Optimize queries to retrieve only the necessary data, minimizing the memory footprint.
- Monitor memory usage and performance metrics to identify potential memory bottlenecks.

4. **Network Bottlenecks**:

Causes:

- Insufficient network bandwidth or network congestion.
- Latency issues caused by network round-trips.
- Inefficient network protocols or communication overhead.

Prevention:

- Ensure adequate network bandwidth and minimize network congestion by optimizing data transfer rates and network infrastructure.
- Minimize network round-trips through query optimization, result set reduction, or caching.
- Utilize efficient network protocols and compression techniques to minimize communication overhead.

5. **Locking and Concurrency Bottlenecks**:

Causes:

- Contentions and conflicts arising from concurrent transactions accessing the same data.
- Inefficient locking mechanisms or lock management strategies.
- High levels of concurrency overwhelming the system's ability to handle simultaneous requests.

Prevention:

- Utilize appropriate concurrency control mechanisms, such as multi-version concurrency control (MVCC) or optimistic locking.
- Optimize transaction isolation levels to balance data consistency and concurrency requirements.
- Implement fine-grained locking to minimize contention and maximize parallelism.
- Regularly monitor and tune concurrency settings based on workload patterns and performance metrics.

Performance bottlenecks in database systems can significantly impact overall system efficiency and responsiveness. By understanding the common causes of these bottlenecks and implementing preventive strategies, organizations can optimize database performance and ensure a smooth user experience. Whether it involves addressing disk I/O issues, optimizing CPU utilization, managing memory efficiently, mitigating network bottlenecks, or optimizing locking and concurrency, proactive measures can be taken to identify, prevent, and resolve performance bottlenecks, leading to enhanced database performance and improved application performance overall.

TECHNIQUES FOR MEASURING AND MONITORING DATABASE PERFORMANCE:

Measuring and monitoring database performance is essential for ensuring optimal system functionality and user experience. By employing effective techniques, organizations can proactively identify performance bottlenecks, track key performance indicators, and make informed decisions to optimize database performance. This section explores various techniques for measuring and monitoring database performance and provides rule-of-thumb performance thresholds as guidelines for assessing system health and identifying potential issues.

PERFORMANCE METRICS AND INDICATORS:

To measure and monitor database performance effectively, it is crucial to track key performance metrics and indicators. These include several metrics that highly depend on the use case and expectations from the computing platform:

a. **Response Time**: The time taken to complete a database operation. Unacceptable threshold: > 1 second for interactive operations, > 5 seconds for batch operations (can vary based on specific requirements).

b. **Throughput**: The rate at which the database processes transactions or queries. Unacceptable threshold: < 100 transactions/queries per second (can vary based on workload and system capacity).

c. **CPU Utilization**: The percentage of CPU time spent on database-related tasks. Unacceptable threshold: > 70-80% (can vary based on system capacity and other workloads).

d. **Memory Usage**: The amount of memory utilized by the database system. Unacceptable threshold: > 80-90% of available memory (can vary based on system requirements and memory management strategies).

e. **Disk I/O**: The rate of data read from and written to disk. Unacceptable threshold: > 10-20 ms disk latency for random I/O, > 1-5 ms for sequential I/O (can vary based on storage technology and disk performance characteristics).

f. **Lock Contention**: The occurrence of locks being held for prolonged periods, causing delays in concurrent transactions. Unacceptable threshold: Lock wait time > 1-2 seconds (can vary based on concurrency requirements and system workload).

PERFORMANCE THRESHOLDS:

While rule-of-thumb performance thresholds can serve as guidelines, it's important to note that they can vary depending on factors such as system requirements, workload characteristics, hardware capabilities, and specific business needs. Organizations should consider these thresholds as starting points and adjust them based on their unique circumstances and performance goals.

a. **Establishing Baselines:** Before setting performance thresholds, establish baselines by monitoring and collecting performance data under normal operating conditions. Baselines provide reference points for identifying deviations from expected performance.

b. **Contextual Considerations**: Tailor performance thresholds to the specific requirements of the application and user expectations. For example, an e-commerce platform with a high volume of transactions may have more stringent performance requirements than an internal reporting system.

c. **Continuous Monitoring and Fine-Tuning**: Regularly review and reassess performance thresholds as system requirements evolve, work load patterns change, or hardware/software upgrades are implemented. Continuous monitoring and fine-tuning ensure that performance thresholds remain relevant and aligned with the evolving needs of the organization.

Measuring and monitoring database performance is vital for maintaining optimal system performance and identifying potential issues. By tracking key performance metrics and employing effective monitoring techniques, organizations can proactively address performance bottlenecks, optimize resource utilization, and deliver a seamless user experience. While rule-of-thumb performance thresholds provide valuable guidelines, they should be adjusted based on contextual considerations and regularly reviewed to ensure they align with evolving system requirements. With comprehensive performance monitoring and thoughtful threshold setting, organizations can stay ahead of performance challenges and take proactive measures to optimize database performance.

MONITORING TOOLS AND TECHNIQUES:

To measure and monitor these performance metrics, organizations can employ various tools and techniques:

a. **Performance Monitoring Tools**: Utilize specialized monitoring tools, such as database-specific monitoring solutions or system-level performance monitoring tools, to collect real-time performance data. Examples include Oracle Enterprise Manager, SQL Server Profiler, and Prometheus.

b. **Database Logs and Event Tracing**: Analyze database logs and event traces to identify performance issues and anomalies. Look for error messages, slow query logs, or event logs indicating resource contention or bottlenecks.

c. **Performance Counters**: Utilize performance counters provided by the database management system (DBMS) or operating system to track key performance metrics. These counters can be accessed through system APIs or monitoring interfaces.

d. **Automated Alerting**: Implement automated alerting mechanisms to notify administrators when performance metrics breach predefined thresholds. Alerts can be configured based on specific metrics and severity levels.

CHAPTER 3. DATABASE DESIGN OPTIMIZATION

Best Practices for Applying the Principles of Efficient Database Schema Design

PRINCIPLES OF EFFICIENT DATABASE SCHEMA DESIGN

Efficient database schema design is a fundamental aspect of database optimization, as it directly influences data integrity, query performance, and overall system efficiency. By adhering to best practices and principles of database schema design, organizations can create robust and scalable databases that enhance data retrieval, minimize redundancy, and ensure optimal performance. This section explores the key best practices for applying the principles of efficient database schema design, providing guidelines for creating well-structured and high-performing database schemas.

1. **Normalize Database Schemas**:

Normalization is a fundamental principle of efficient database schema design. It involves decomposing data into logical entities, eliminating redundancy, and ensuring data integrity through well-defined relationships. Key normalization forms, which are: First Normal Form (1NF), Second Normal Form (2NF), and Third Normal Form (3NF), help minimize data duplication, improve query performance, and maintain data consistency. Applying normalization principles prevents data anomalies and simplifies database maintenance and updates.

2. **Understand and Model the Business Requirements**:

Before designing a database schema, it is essential to have a thorough understanding of the business requirements and data entities involved. Identify the entities, relationships, and attributes required to represent the business domain accurately. Collaborate with stakeholders, subject matter experts, and application users to capture and analyze the data requirements comprehensively. This ensures that the database schema aligns closely with the business needs, facilitating efficient data retrieval and manipulation.

3. **Establish Appropriate Primary and Foreign Key Relationships**:

Defining proper primary key and foreign key relationships is vital for maintaining data integrity and supporting efficient data retrieval. Identify the primary key attributes for each table to uniquely identify records. Establish appropriate foreign key relationships to establish connections between related tables. Consistently enforce referential integrity by ensuring that foreign keys reference valid primary keys. Well-defined relationships enable efficient query execution, join operations, and enforce data consistency.

4. **Avoid Overuse of Denormalization**:

While denormalization can enhance query performance in specific scenarios, it should be used judiciously. Denormalization involves intentionally introducing redundancy in the schema to optimize specific query patterns. However, overusing denormalization can lead to data inconsistencies and maintenance complexities. Evaluate the trade-offs carefully and apply denormalization sparingly, considering factors such as query patterns, performance requirements, and data update frequency.

5. **Optimize Data Types and Field Sizes**:

Choosing appropriate data types and field sizes for schema attributes is crucial for efficient storage and query execution. Select data types that accurately represent the data and minimize storage requirements. Avoid using oversized data types that waste storage space or undersized types that may cause data truncation. Consider the maximum expected data size while setting field lengths to optimize storage and memory utilization.

6. **Implement Indexing Strategically**:

Indexing plays a crucial role in optimizing query performance. Identify the frequently queried attributes and establish indexes on those columns. Indexes facilitate rapid data retrieval by reducing the need for full table scans. However, excessive indexes can negatively impact insert and update operations. Selectively apply indexes based on query patterns, performance requirements, and the balance between read and write operations.

7. **Plan for Future Scalability**:

Anticipate future growth and scalability requirements while designing the database schema. Consider factors such as increased data volumes, user concurrency, and potential system expansions. Design schemas that support horizontal scalability through techniques like data partitioning or sharding. Establish a schema that can evolve with the organization's needs, reducing the need for major schema redesigns in the future.

8. **Regularly Perform Schema Reviews and Refactoring**:

Database schema design is an iterative process. Regularly review the schema design and perform refactoring to optimize performance and maintain data integrity. As the application and data requirements evolve, periodically assess the schema for potential enhancements or optimizations. Identify opportunities for refactoring, such as splitting tables, removing unused columns, or reorganizing relationships, to streamline the schema and improve performance.

9. **Consider Performance and Security Implications**:

Efficient database schema design should also consider performance and security implications. Take into account the anticipated workload, query patterns, and access patterns to optimize performance. Consider data access controls, encryption, and data privacy requirements to ensure data security and compliance with regulations.

10. **Document and Communicate the Schema Design**:

Document the database schema design comprehensively, including entity-relationship diagrams, data dictionaries, and any design decisions or considerations. Clear documentation facilitates better understanding and communication among stakeholders, developers, and administrators. It also serves as a reference for future maintenance and modifications.

Efficient database schema design is crucial for optimizing database performance, ensuring data integrity, and facilitating scalability. By following best practices such as normalization, understanding business requirements, establishing proper relationships, and optimizing data types and indexes, organizations can create high-performing and maintainable database schemas. Regular reviews, refactoring, and considering performance and security implications further contribute to the effectiveness of the schema design. Applying these best practices fosters a solid foundation for efficient data retrieval, robust application performance, and streamlined database management.

SCHEMA ADAPTATION AND SCALING USING OBJECT-RELATIONAL MAPPING

An Object-Relational Mapping (ORM) tool plays a crucial role in scaling and adapting to refactoring database schemas by providing a layer of abstraction between the application code and the underlying database. Here's how an ORM tool helps in these aspects:

1. **Scalability**: When scaling a database system, changes to the schema structure are often required to accommodate increased data volumes or changes in application requirements. An ORM tool simplifies the process of scaling by abstracting the database interactions through object-oriented programming paradigms. This allows developers to focus on the application logic without worrying about low-level database details.

➢**Schema Migrations**: ORM tools often provide built-in mechanisms for managing schema migrations. These tools assist in modifying the database schema while maintaining data integrity and minimizing downtime. They can automatically generate and execute SQL scripts to migrate the database schema, adding or modifying tables, columns, indexes, and relationships.

➢**Data Access Layer**: An ORM tool abstracts the database access layer, providing an interface that shields the application code from underlying database changes. As the database schema evolves during scaling, the ORM tool handles the mapping between the new schema and the application objects seamlessly, reducing the impact of schema changes on the code-base.

2. **Adaptability to Refactoring**:

Refactoring involves restructuring the code-base to improve its design, maintainability, and performance. Refactoring often impacts the database schema as well. Here's how an ORM tool assists in adapting to database schema refactoring:

- **Object-Relational Mapping**: An ORM tool maps database tables to application objects, allowing developers to work with familiar object-oriented constructs. When refactoring the database schema, the ORM tool's mapping configuration can be updated to reflect the changes. This enables the application code to adapt to the new schema seamlessly without requiring extensive modifications.

- **Automatic Query Generation**: ORM tools generate SQL queries based on the defined object mappings. As the database schema is refactored, the ORM tool dynamically adjusts the query generation process to align with the new schema structure. Developers can continue to work with the existing object-oriented code-base while the ORM tool handles the translation of object-oriented operations into database-specific queries.

- **Data Migration**: During schema refactoring, data may need to be transformed or migrated to conform to the new schema structure. ORM tools often provide utilities and mechanisms to facilitate data migration, allowing developers to write data transformation scripts and automate the process of migrating data from the old schema to the new schema.

- **Schema Versioning**: ORM tools typically include features for managing schema versions and migrations. This enables developers to track and manage the evolution of the database schema over time, ensuring that the application remains compatible with the schema changes during refactoring.

Overall, an ORM tool abstracts the complexities of the underlying database, providing a flexible and adaptable interface for application developers. It simplifies the process of scaling and adapting to refactoring database schemas by automating schema migrations, managing object-relational mapping, and facilitating data migration. This allows developers to focus on the application logic and design, promoting agility, maintainability, and scalability in the face of evolving database schemas.

COMMON PITFALLS WITH SCHEMA DESIGN

Improper schema design can lead to various issues, including performance degradation, data inconsistencies, and maintenance challenges. By being aware of common pitfalls and adopting preventive measures, organizations can avoid these problems. Here are some common pitfalls of improper schema design and ways to avoid them:

1. **Denormalization Without Justification**: Overuse of denormalization, introducing redundant data, can lead to data inconsistencies, increased storage requirements,

and maintenance complexities. Prevention: Employ denormalization judiciously, considering specific performance requirements and query patterns. Document the rationale behind denormalization decisions and regularly review the denormalized schema to ensure data integrity.

2. **Lack of Proper Indexing**: Insufficient or improper indexing can result in slow query performance and excessive resource consumption. Prevention: Analyze query patterns, identify frequently accessed columns, and establish appropriate indexes. Regularly monitor and tune indexes based on query performance metrics. Be cautious not to create excessive indexes, as they can impact insert and update operations negatively.

3. **Failure to Enforce Data Integrity**: Inadequate enforcement of data integrity constraints can lead to data inconsistencies and compromise data quality. Prevention: Define and enforce primary key constraints, foreign key relationships, and other integrity constraints. Utilize database features like referential integrity checks and cascading updates/deletes to maintain data consistency. Regularly validate and verify data integrity through data quality checks and audits.

4. **Inefficient Data Types and Field Sizes**: Improper selection of data types and field sizes can result in wasted storage space, performance degradation, or data truncation issues. Prevention: Choose appropriate data types that accurately represent the data and optimize storage. Consider the maximum expected data size when setting field lengths. Regularly review and adjust data types and field sizes based on actual data usage and growth patterns.

5. **Failure to Plan for Scalability**: Neglecting scalability considerations during schema design can lead to difficulties in accommodating future growth and performance requirements. Prevention: Anticipate future data volumes, user concurrency, and system expansions. Design schemas that support horizontal scalability through techniques like data partitioning or sharding. Establish a schema that can evolve with the organization's needs, reducing the need for major schema redesigns.

6. **Insufficient Documentation and Communication**: Inadequate documentation and communication about the schema design can hinder understanding, collaboration, and maintenance efforts. Prevention: Document the database schema comprehensively, including entity-relationship diagrams, data dictionaries, and design decisions. Clearly communicate the schema design and its rationale to stakeholders, developers, and administrators. Regularly update the documentation as the schema evolves.

7. **Ignoring Performance Testing and Optimization**: Neglecting performance testing and optimization can result in sub-optimal query performance, resource inefficiencies, and scalability issues. Prevention: Perform thorough performance testing to identify bottlenecks and optimize query execution plans. Utilize query profiling, monitoring tools, and performance benchmarks to fine-tune the schema

and query performance. Regularly monitor performance metrics, identify performance degradation, and proactively address performance issues.

By avoiding these common pitfalls and adopting best practices in schema design, organizations can create efficient, scalable, and maintainable database schemas. Regular reviews, refactoring, and performance optimization should be incorporated as ongoing practices to ensure the schema design aligns with evolving requirements and maintains optimal performance.

NORMALIZATION AND DENORMALIZATION: TECHNIQUES AND BEST PRACTICES

Normalization and denormalization are two complementary approaches in database design aimed at achieving efficient, scalable, and maintainable database schemas. Normalization involves decomposing data into logical entities to eliminate redundancy and maintain data integrity, while denormalization selectively introduces redundancy to improve query performance. This section explores the concepts, techniques, and best practices of normalization and denormalization, providing insights into when and how to apply each approach effectively.

NORMALIZATION:

Normalization is the process of organizing data into logical entities, or tables, to minimize redundancy and eliminate data anomalies. It follows a set of normalization forms, each building upon the previous one, to achieve data integrity. Here are the key normalization forms:

FIRST NORMAL FORM (1NF):

In 1NF, data is organized into tables with atomic (indivisible) values, and each attribute has a single value for each record. This eliminates repeating groups and ensures that each column contains only one piece of information.

First Normal Form (1NF) is the first step in the process of database normalization. It requires that a relation (table) have no repeating groups and that each attribute hold only atomic (indivisible) values. Here's a simple example of transforming a table into 1NF:

Original Table: Customer Orders

OrderID	Customer	Items
101	Abel	Item1, Item2
102	Bill	Item2, Item3, Item4
103	Catherine	Item1

In the original table, the "Items" column contains a repeating group (comma-separated values) for each order. This violates 1NF. To bring it into 1NF, we need to break down the repeating group into separate rows:

1NF Transformed Table: Customer Orders

OrderID	Customer	Item
101	Abel	Item1
101	Abel	Item2
102	Bill	Item2
102	Bill	Item3
102	Bill	Item4
103	Catherine	Item1

Now, each order has its own row, and the repeating items have been split into separate rows. This adheres to 1NF by eliminating repeating groups and ensuring atomicity for each attribute.

Benefits of 1NF:

➢**Data Consistency:** 1NF ensures that each column holds only atomic values, reducing the chance of data inconsistencies and anomalies.

➢**Query Flexibility:** Data stored in 1NF can be queried more easily, as it's structured in a way that aligns with relational database principles.

➢**Efficient Updates:** Storing data in 1NF simplifies updates and modifications, as each attribute represents a single value.

However, note that 1NF does not eliminate all anomalies; further normalization steps (2NF, 3NF, etc.) may be necessary to achieve higher levels of data integrity and optimization.

SECOND NORMAL FORM (2NF):

2NF builds on 1NF by eliminating partial dependencies. It requires that every non-key attribute is functionally dependent on the entire primary key, not just part of it. This form ensures that data is organized efficiently, avoiding redundancy and potential update anomalies.

Second Normal Form (2NF) is the next step in database normalization. It builds upon First Normal Form (1NF) by requiring that all non-key attributes be fully functionally dependent on the entire primary key. Here's an example of transforming a table into 2NF:

Original Table: Customer Orders

OrderID	Customer	Item	Category
101	Abel	Item1	Electronics
101	Abel	Item2	Electronics
102	Bill	Item2	Furniture
102	Bill	Item3	Furniture
102	Bill	Item4	Furniture
103	Catherine	Item1	Electronics

In the original table, the "Category" attribute depends only on the "Item" attribute, and not directly on the entire primary key (OrderID, Customer). This violates 2NF. To bring it into 2NF, we need to create a separate table for the categories:

2NF Transformed Table: Customer Orders

OrderID	Customer	Item
101	Abel	Item1
101	Abel	Item2
102	Bill	Item2
102	Bill	Item3
102	Bill	Item4
103	Catherine	Item1

Categories Table:

Item	Category
Item1	Electronics
Item2	Electronics
Item3	Furniture
Item4	Furniture

In this transformed structure, the "Category" attribute has been removed from the "Customer Orders" table and placed in a separate "Categories" table. This adheres to 2NF because the "Category" attribute is now fully functionally dependent on the primary key of the "Categories" table (Item).

Benefits of 2NF:

- **Elimination of Partial Dependencies:** 2NF eliminates partial dependencies, ensuring that attributes are dependent on the entire primary key.
- **Reduced Data Duplication:** Storing categories in a separate table eliminates redundant data and improves data consistency.
- **Simplified Updates:** Separating data into related tables simplifies updates and modifications, as changes need to be made in fewer places.

2NF, while an important step, might not be the final normalization step depending on the complexity of the data and the relationships. Further normalization (3NF, BCNF, etc.) can provide higher levels of data integrity and optimization.

THIRD NORMAL FORM (3NF):

3NF further refines the schema by removing transitive dependencies. It ensures that non-key attributes are not dependent on other non-key attributes. By eliminating transitive dependencies, 3NF minimizes redundancy and improves data integrity.

Third Normal Form (3NF) is the next step in database normalization. It builds upon Second Normal Form (2NF) by removing transitive dependencies, ensuring that non-key attributes are dependent only on the primary key. Here's an example of transforming a table into 3NF:

Original Table: Customer Orders

OrderID	Customer	Item	Category	Category Description
101	Abel	Item1	Electronics	Electronics Store
101	Abel	Item2	Electronics	Electronics Store
102	Bill	Item2	Furniture	Furniture Store
102	Bill	Item3	Furniture	Furniture Store
102	Bill	Item4	Furniture	Furniture Store
103	Catherine	Item1	Electronics	Electronics Store

In the original table, there is a transitive dependency between "Category Description" and "Category." The "Category Description" depends on the "Category," which in turn depends on the "Item." This violates 3NF. To bring it into 3NF, we need to create separate tables for categories and category descriptions:

3NF Transformed Table: Customer Orders

OrderID	Customer	Item	Category
101	Abel	Item1	Electronics
101	Abel	Item2	Electronics
102	Bill	Item2	Furniture
102	Bill	Item3	Furniture
102	Bill	Item4	Furniture
103	Catherine	Item1	Electronics

Categories Table:

Category	Category Description
Electronics	Electronics Store
Furniture	Furniture Store

In this transformed structure, the "Category Description" has been removed from the "Customer Orders" table and placed in a separate "Categories" table. This adheres to 3NF by eliminating the transitive dependency.

Benefits of 3NF:

- **Elimination of Transitive Dependencies:** 3NF removes transitive dependencies, ensuring that attributes are only dependent on the primary key.
- **Reduction of Data Duplication:** Storing category descriptions in a separate table eliminates data redundancy.
- **Improved Data Integrity:** 3NF ensures that data is stored logically and prevents anomalies due to transitive dependencies.

It's important to note that while 3NF addresses many potential data integrity issues, further normalization (e.g., Boyce-Codd Normal Form or Fourth Normal Form) may be necessary for more complex scenarios. In practice, the added complexity, the extra query overhead (using joins) along with the table cardinality to assess the trade-offs (explained below) with this level and further normalization.

BOYCE-CODD NORMAL FORM (BCNF)

Boyce-Codd Normal Form (BCNF) is a higher level of database normalization that builds upon Third Normal Form (3NF) by addressing further anomalies related to functional dependencies. To achieve BCNF, each non-key attribute must be fully functionally dependent on the primary key, and there should be no non-trivial dependencies among non-key attributes. Here's an example of transforming a table into BCNF:

Original Table: Student Courses

StudentID	CourseID	Course Name	Professor	Professor Office
101	CS901	Adv CS	Hendricks	Room 202
102	ASTR303	Cosmology	Dickens	Room 105
101	ASTR303	Cosmology	Dickens	Room 105
103	CS901	Adv CS	Hendricks	Room 202

In the original table, there are functional dependencies that violate BCNF. Both "Course Name" and "Professor" are functionally dependent on "CourseID," but they are also functionally dependent on each other through "CourseID." This redundancy leads to anomalies. To bring it into BCNF, we need to split the table into separate tables for courses and professors:

BCNF Transformed Table: Student Courses

StudentID	CourseID
101	CS901
102	ASTR303
101	ASTR303
103	CS901

Courses Table:

CourseID	Course Name
CS901	Adv CS
ASTR303	Cosmology

Professors Table:

CourseID	Professor	Professor Office
CS901	Hendricks	Room 202
ASTR303	Dickens	Room 105

In this transformed structure, we've separated the "Course Name" and "Professor" information into independent tables. This adheres to BCNF by eliminating non-trivial functional dependencies between non-key attributes.

Benefits of BCNF:

- **Elimination of Anomalies:** BCNF eliminates anomalies arising from non-trivial functional dependencies, such as insertion, update, and deletion anomalies.
- **Data Consistency:** BCNF ensures that data is stored logically and consistently, improving data quality.
- **Reduced Data Redundancy:** By removing redundant information, BCNF minimizes data duplication and storage inefficiencies.

Achieving BCNF is important for maintaining high levels of data integrity and preventing various anomalies that can arise during database operations.

While Boyce-Codd Normal Form (BCNF) is a powerful and desirable level of database normalization for ensuring data integrity and minimizing anomalies, it's important to recognize that achieving BCNF can come with trade-offs and potential drawbacks:

1. Complexity of Design: Transforming a schema into BCNF might require breaking down tables into smaller, more normalized forms. This can lead to a more complex database design with a larger number of tables, potentially making the schema harder to understand and manage.

2. Increased Joins: Splitting tables into smaller components in BCNF can lead to increased need for JOIN operations in queries to retrieve related information. While JOINS are a fundamental aspect of relational databases, excessive use of complex JOINs can impact query performance.

3. Query Performance: The normalized structure in BCNF might result in more JOIN operations to retrieve the required data, which can affect query performance. Depending on the complexity of the schema and the query workload, performance could be impacted, especially if the database is not properly indexed.

4. Redundant Data Access: In highly normalized schemas, it's possible that related data ends up stored in different tables. This could lead to additional data access steps, which might not be as efficient as retrieving denormalized data from a single table.

5. Storage Overhead: Normalization can lead to more tables and potentially smaller rows due to the removal of redundant data. However, the overhead of maintaining multiple tables, indexes, and relationships might lead to increased storage requirements.

6. Update Complexity: In BCNF, related data might be stored in different tables. This can lead to more complex update operations, as changes need to be propagated across multiple tables to maintain data consistency.

7. Application Complexity: Applications interacting with a BCNF-normalized database might need to handle complex queries, JOIN operations, and data retrieval strategies. This can lead to more complex application code, potentially impacting development time and maintenance.

8. Read-Heavy vs. Write-Heavy Workloads: BCNF might be more suitable for read-heavy workloads, as it optimizes data integrity and consistency. However, for write-heavy workloads, the increased complexity of updates due to normalization might impact performance.

It's important to note that while BCNF can introduce these trade-offs, the decision to normalize to BCNF or not depends on various factors, including the nature of the data, the application's query and update patterns, and performance considerations. Striking the right balance between normalization and performance is essential for designing a database schema that meets both data integrity and performance goals.

BEST PRACTICES FOR NORMALIZATION:

- Understand the business requirements and data dependencies before starting the normalization process.
- Identify and define entities, attributes, and relationships accurately.
- Follow normalization forms progressively, ensuring each form's requirements are met before moving to the next.
- Regularly review and revise the schema based on changing requirements to maintain data integrity and minimize redundancy.
- Use referential integrity constraints, such as primary key and foreign key relationships, to enforce data consistency.
- Balance normalization and query performance considerations, as highly normalized schemas may require complex joins and impact query performance.

DENORMALIZATION:

Denormalization involves selectively introducing redundancy into a normalized schema to improve query performance. It aims to optimize data retrieval by reducing the need for complex joins and improving data locality. Denormalization is typically used in scenarios where read performance is deemed critical, such as data warehouses or reporting systems.

COMMON DENORMALIZATION TECHNIQUES:

1. Flattening: Combining related tables into a single table to reduce join operations and simplify queries. This reduces the need for complex joins and improves query performance.

2. Prejoining: Storing precomputed results of join operations as separate tables. This eliminates the need for frequent join calculations and speeds up query execution.

3. Introducing Redundant Columns: Duplicating selected data from related tables into a single table. This eliminates the need for joins and reduces the number of tables involved in query processing.

BEST PRACTICES FOR DENORMALIZATION:

- Analyze query patterns and identify performance-limiting queries that can benefit from denormalization.

- Consider denormalization as an optimization technique, applied selectively to specific tables or attributes, rather than denormalizing the entire schema.

- Document and communicate denormalization decisions to ensure understanding and maintainability.

- Implement mechanisms, such as triggers or stored procedures, to ensure data consistency when denormalizing.

- Regularly review and reassess denormalization choices based on changing requirements and performance monitoring.

NORMALIZATION VS. DENORMALIZATION:

Choosing between normalization and denormalization depends on the specific requirements and trade-offs of the application. While normalization ensures data integrity and reduces redundancy, it may result in complex joins and slower query performance. Denormalization improves query performance but introduces redundancy, potentially impacting maintenance and data consistency. Here are some considerations when deciding between normalization and denormalization:

1. Read vs. Write Performance: If read performance is a top priority and the data is relatively stable, denormalization can be beneficial. On the other hand, if write performance and data integrity are deemed critical, normalization is preferred.
2. Data Updates: Denormalization can complicate data updates, as changes may need to be propagated across multiple redundant columns or tables. Normalization simplifies updates by maintaining data in a single location.
3. Query Complexity: Highly normalized schemas may require complex joins to retrieve data, impacting query performance. Denormalization can simplify queries and reduce join operations, improving query response time.
4. Application Requirements: Consider the nature of the application and the types of queries performed. OLTP (Online Transaction Processing) systems usually benefit from normalization, while OLAP (Online Analytical Processing) or reporting systems often benefit from denormalization.
5. Maintenance and Extensibility: Normalized schemas are generally easier to maintain and extend, as changes are localized to specific tables. Denormalized schemas require careful management to ensure data consistency and may be more challenging to modify or add new features.
6. Data Consistency and Integrity: Normalization enforces data consistency through referential integrity constraints. Denormalization requires additional measures, such as triggers or stored procedures, to ensure data integrity.

Normalization and denormalization are complementary techniques in database schema design. Normalization aims to eliminate redundancy and maintain data integrity, while denormalization selectively introduces redundancy to improve query performance. Best practices involve understanding business requirements, applying normalization forms progressively, enforcing data integrity, and considering trade-offs when deciding to denormalize. By carefully evaluating the specific needs of an application and striking a balance between normalization and denormalization, organizations can design efficient, scalable, and maintainable database schemas that meet performance and data integrity requirements.

INDEXING STRATEGIES FOR IMPROVED QUERY PERFORMANCE

Indexing plays a crucial role in improving query performance by facilitating rapid data retrieval and reducing the need for full table scans. Implementing effective indexing strategies can significantly enhance the efficiency of database systems. Here are some key indexing strategies to consider for improved query performance:

1. Identify High-Frequency Query Columns: Analyze the queries executed against the database and identify the frequently accessed columns. These columns are good

candidates for indexing as they are often used in filtering or joining conditions. By indexing these columns, the database can quickly locate the relevant data, improving query execution speed.

2. Establish Primary Key and Unique Indexes: Primary key and unique constraint columns should be indexed automatically by most database systems. These indexes ensure data integrity and support fast lookup operations when performing primary key or unique value-based searches. Additionally, they enhance join performance when these columns are used as foreign keys in related tables.

3. Utilize Composite Indexes: Composite indexes involve multiple columns and are useful when queries involve filtering or sorting based on multiple columns simultaneously. By creating an index on a combination of columns frequently used in queries, the database can avoid performing separate index look-ups for each individual column, improving overall query performance.

4. Consider Indexing Foreign Key Columns: Foreign key columns are often used in join operations between related tables. Indexing these columns can enhance the efficiency of such join operations, resulting in faster query execution. It is especially important when joining large tables or executing complex queries involving multiple tables.

5. Analyze Query Patterns for Covering Indexes: Covering indexes are indexes that include all the columns required by a query, allowing the database to retrieve the necessary data directly from the index without referring to the actual table. Analyze query patterns to identify opportunities for covering indexes, which can eliminate the need for table look-ups, significantly improving query performance.

6. Regularly Monitor and Tune Indexes: Monitoring and analyzing index usage, query performance, and system resources are vital for maintaining optimal indexing strategies. Regularly review the performance of queries, identify slow-running queries, and consider index modifications or additions to address performance bottlenecks. Avoid maintaining unnecessary or redundant indexes, as they can introduce overhead and impact overall system performance.

7. Be Mindful of Index Overhead: While indexes enhance query performance, they come with a trade-off in terms of storage space and maintenance overhead. Carefully assess the trade-offs and consider the impact of indexes on data modification operations (e.g., inserts, updates, and deletes). Excessive or inappropriate indexing can lead to increased disk space usage, slower data modifications, and increased overhead during data maintenance.

8. Leverage Indexing Tools and Expertise: Database management systems offer tools and utilities to analyze query performance, identify missing or underutilized indexes, and suggest index optimizations. Utilize these tools to gain insights into

indexing strategies and optimize query performance. Consider consulting with database experts or database administrators who have expertise in performance tuning and indexing.

Monitoring high-frequency query columns and index usage is crucial for optimizing database performance. This helps you identify which columns are frequently accessed and whether your indexes are effectively supporting query performance. Here's how you can monitor these aspects that are detailed in a proceeding chapter:

1. Query Profiling: Enable query profiling in your database system. This feature records information about each query's execution, including the columns accessed, indexes used, and execution time. Analyzing query profiles can reveal which columns and indexes are being utilized most frequently.

2. Database Performance Tools: Utilize performance monitoring and management tools provided by your database system. These tools often offer insights into query execution plans, including which indexes are used, helping you understand the impact of high-frequency queries.

3. Database Logs: Review database logs to identify which columns and indexes are being accessed frequently. These logs might provide insights into the most active parts of your database.

4. Index Usage Statistics: Most database systems provide statistics on index usage. Monitor these statistics to determine which indexes are heavily used and which are not. This information helps you decide whether to optimize, add, or remove certain indexes.

5. Database Queries Analysis: Regularly analyze your database queries to identify patterns. Look for queries that frequently join specific tables or filter on particular columns. These patterns suggest high-frequency query columns.

6. Profiling Tools and Libraries: Many database libraries and ORM (Object-Relational Mapping) tools offer built-in profiling capabilities. These tools can provide insights into which columns are frequently accessed when fetching data through the application.

7. User Behavior Analysis: Understand user behavior and application usage patterns. Frequently accessed parts of your application often correlate with high-frequency query columns. Analyzing user interactions can help identify the columns that require indexing.

8. Data Analytics: If you have a data analytics platform, you can analyze query logs or traffic patterns to identify frequently queried columns and their corresponding indexes.

9. Index Health Checks: Regularly evaluate the health of your indexes. Unused or redundant indexes can slow down data modifications and increase storage requirements. Removing or optimizing such indexes improves overall performance.

10. Performance Testing: During performance testing, simulate real-world workloads and monitor query and index usage patterns. This can reveal how high-frequency query columns and indexes behave under different conditions.

By diligently monitoring high-frequency query columns and index usage, you can gain insights into which parts of your database are critical for performance and which indexes might need optimization. This information guides your indexing strategies, helps you make informed decisions about index creation and maintenance, and ultimately leads to a well-optimized database system.

Indexing is a powerful technique for improving query performance in database systems. By identifying key query columns, establishing primary key and unique indexes, utilizing composite and covering indexes, and regularly monitoring and tuning indexes, organizations can enhance the efficiency of data retrieval and overall system performance. However, it is important to strike a balance between indexing and the associated overhead to ensure optimal performance across different types of queries and data manipulation operations.

There are several types of indexes available in database systems, each with its own strengths and weaknesses. Understanding these index types and their appropriate use cases can help optimize query performance based on specific requirements. Here are some commonly used types of indexes:

- B-Tree Index: B-Tree (Balanced Tree) indexes are the most common and widely used index type. They are well-suited for range-based queries and equality searches. B-Tree indexes provide efficient data retrieval for both ordered and unordered data. They maintain sorted key-value pairs in a tree structure, enabling fast lookup and range-based operations. B-Tree indexes work well for scenarios where the data is frequently updated, as they dynamically balance and adjust to changes. However, they may not be as efficient for queries that require complex pattern matching or partial string searches.

- Bitmap Index: Bitmap indexes are particularly useful for columns with low cardinality, where the number of distinct values is relatively small. They represent each distinct value as a bit vector or bitmap, with each bit indicating the presence or absence of the value in the corresponding row. Bitmap indexes are efficient for filtering operations, especially when multiple conditions are combined using logical AND or OR operators. However, bitmap indexes can be memory-intensive and may not be suitable for columns with high cardinality or frequently changing data.

- Hash Index: Hash indexes use a hash function to map keys to specific locations, providing fast equality searches. They are efficient for exact match lookups but do not support range-based queries or sorting. Hash indexes work well for columns with high cardinality and a uniform distribution of values. However, they are less effective for queries involving inequality comparisons or wildcard matching. Hash indexes can also be sensitive to hash collisions, which may impact performance.

➢Clustered Index: A clustered index determines the physical order of data in a table. It defines the order of rows based on the indexed column(s), resulting in efficient retrieval of ordered data. Clustered indexes are particularly beneficial for queries that retrieve a range of consecutive values or perform sorting based on the indexed column(s). However, maintaining a clustered index can be resource-intensive, especially when inserting or updating data, as it requires physically reorganizing the data.

➢Non-Clustered Index: Non-clustered indexes store a copy of the indexed column(s) along with a pointer to the actual data. They provide efficient data retrieval for specific columns, supporting both equality and range-based searches. Non-clustered indexes are useful for queries that involve join operations, sorting, or filtering based on specific columns. They are less resource-intensive than clustered indexes but may result in additional disk space usage due to storing redundant copies of the indexed columns.

➢Full-Text Index: Full-Text indexes are designed to enhance text-based search capabilities in large text fields or documents. They enable efficient pattern matching, linguistic analysis, and relevance ranking. Full-Text indexes are suitable for scenarios where complex text search operations are required, such as searching for keywords, phrases, or performing fuzzy searches. They may have higher storage requirements compared to other index types due to the need to index textual content comprehensively.

Choosing the appropriate index type depends on the specific requirements and characteristics of the data and queries. Consider the following factors when deciding which index type to use:

➢Data Distribution: Consider the cardinality and distribution of values in the indexed column(s). Some index types perform better with low cardinality, while others excel with high cardinality.

➢Query Patterns: Analyze the types of queries executed against the data. Determine if the queries involve range-based searches, equality searches, or complex text searches to select the index type that aligns with the query patterns.

➢Data Modification Frequency: Assess the frequency of data modifications (inserts, updates, deletes). Consider the impact of index maintenance overhead on data modification operations. Some index types, like B-Tree indexes, handle frequent data modifications more efficiently than others.

➢Data Size and Memory Constraints: Consider the size of the indexed data and available memory resources. Bitmap indexes, for example, can be memory-intensive, so the system's memory capacity should be taken into account.

➢Join and Sorting Operations: Evaluate whether the queries involve join operations, sorting, or filtering based on specific columns. Non-clustered indexes and clustered indexes can be beneficial in such scenarios.

➢Text Search Requirements: If the queries involve text-based searches, full-text indexes are well-suited to handle complex text search operations efficiently.

It is worth noting that multiple index types can be combined and used together, depending on the specific needs of the application and the queries being executed. It is essential to regularly monitor and evaluate the performance of the indexes, as well as their impact on data modification operations, to ensure optimal index selection and usage.

Choosing the appropriate index type involves considering factors such as data distribution, query patterns, data modification frequency, memory constraints, and specific requirements like text search capabilities. Understanding the strengths and weaknesses of different index types enables effective indexing strategies that optimize query performance and enhance overall database system efficiency.

There are specialized index types included with most major DBMS. It is important to note that the suitability and performance of specialized index types can vary depending on the specific data characteristics, query patterns, and database system being used. Evaluating the data requirements, access patterns, and performance considerations will help determine the appropriate index type for a given scenario.

In addition to the previously mentioned index types, there are a few other specialized index types that can be applied in specific scenarios to improve query performance and address unique requirements. Here are a few examples:

1. Spatial Index: A spatial index is designed to optimize the storage and retrieval of spatial data, such as points, lines, and polygons. It enables efficient spatial queries, such as finding nearby locations or performing geometric operations. Spatial indexes are suitable for applications that involve geospatial data, mapping, or location-based services. By organizing data based on spatial proximity, spatial indexes provide fast access to relevant spatial information.

2. Function-Based Index: A function-based index allows indexing the result of a function or expression rather than just a column. It can be created based on calculated or transformed values of existing columns. Function-based indexes are useful when queries involve complex calculations, string manipulations, or custom functions. By precomputing and indexing the transformed data, function-based indexes can significantly enhance query performance for such scenarios.

3. Reverse Index: A reverse index, also known as an inverted index, is commonly used in full-text search engines. It maps terms or keywords to the documents or records that contain them. Reverse indexes enable fast keyword-based searches, making them ideal for applications that require efficient text search capabilities.

They are commonly used in document management systems, search engines, and content repositories.

4. Partial Index: A partial index is a type of index that includes a subset of rows from a table, based on a defined condition or filter. It indexes only the rows that meet the specified criteria, reducing the index size and improving query performance for specific subsets of data. Partial indexes are useful when queries predominantly access a subset of data, such as active records or frequently accessed rows. They help optimize query execution by excluding unnecessary data from the index.

5. Clustered Columnstore Index: A clustered columnstore index is a specialized index type designed for data warehousing and analytics workloads. It stores and compresses data in columnar format, improving query performance for large-scale analytical queries. Clustered columnstore indexes are well-suited for scenarios that involve high-volume data processing, batch operations, or analytical queries that scan and aggregate large datasets.

When considering these specialized index types, it's important to assess the specific requirements of the application, the nature of the data, and the types of queries being executed. Careful analysis of query patterns, data characteristics, and performance considerations will help determine the suitability and potential benefits of these index types for a given scenario.

It's worth noting that not all database management systems support every index type mentioned here. The availability and implementation details may vary depending on the specific database system being used.

PARTITIONING AND SHARDING APPROACHES: OPTIMIZING SCALABILITY AND PERFORMANCE

In modern database systems, partitioning and sharding are two commonly employed approaches to optimize scalability, improve performance, and handle large-scale data processing. Both techniques involve distributing data across multiple nodes or partitions to leverage parallelism and accommodate increasing data volumes. This section explores partitioning and sharding approaches, discussing their concepts, benefits, implementation strategies, and considerations for effective implementation.

PARTITIONING:

Partitioning, also known as horizontal partitioning, involves dividing a large database into smaller, more manageable subsets called partitions. Each partition contains a subset of data with its own schema, stored on separate storage units or nodes. Partitioning offers several benefits:

1. Scalability: Partitioning allows for distributing data across multiple servers or storage devices, enabling parallel processing and accommodating large data volumes. It enhances scalability by reducing the load on individual nodes and facilitating data distribution.

2. Performance: Partitioning improves query performance by narrowing the scope of data retrieval. Instead of scanning the entire dataset, queries can be directed to specific partitions, reducing the amount of data accessed and improving response times.

3. Manageability: Partitioning enhances manageability by isolating data subsets. It simplifies maintenance operations, such as backups, restores, and data archiving, as these operations can be performed on individual partitions independently.

PARTITIONING STRATEGIES:

RANGE PARTITIONING:

Data is partitioned based on a defined range of values, such as date ranges or numerical ranges. Each partition contains data falling within the specified range. This method particularly effective in improving performance with high cardinality tables.

Range partitioning is a database partitioning technique where data is divided into partitions based on a specified range of values in a chosen column. Each partition contains data within a specific range, making it easier to manage and query large datasets. Here's a simple example of range partitioning in a table:

Scenario: Sales Orders Partitioning

Consider a sales database with a table named "Orders." You want to partition the "Orders" table based on the order date to improve query performance.

Original Table: Orders

OrderID	OrderDate	Customer	TotalAmount
1	2023-01-15	Abel	1000
2	2023-02-20	Bill	1500
3	2023-02-25	Catherine	800
...	...	...	...

You decide to partition the "Orders" table based on the "OrderDate" column. You create partitions for specific date ranges:

➢Partition 1: Orders before 2023-02-01

➢Partition 2: Orders between 2023-02-01 and 2023-03-01

➢Partition 3: Orders after 2023-03-01

After partitioning, the data would be physically organized like this:

Partition 1: Orders before 2023-02-01

OrderID	OrderDate	Customer	TotalAmount
1	2023-01-15	Abel	1000
...	...	...	...

Partition 2: Orders between 2023-02-01 and 2023-03-01

OrderID	OrderDate	Customer	TotalAmount
2	2023-02-20	Bill	1500
3	2023-02-25	Catherine	800
...	...	...	...

Partition 3: Orders after 2023-03-01

OrderID	OrderDate	Customer	TotalAmount
...	...	...	...

Benefits of Range Partitioning:

➢**Improved Query Performance:** Range partitioning allows queries to target specific partitions, reducing the amount of data that needs to be scanned.

➢**Efficient Data Maintenance:** Partitioning enables easier data archiving and maintenance, especially for historical data.

➢**Scalability:** Range partitioning supports horizontal scalability by distributing data across different storage devices or servers.

Considerations:

- ➢Choose a partitioning column that makes sense for your data access patterns.
- ➢Regularly monitor and maintain partitions, especially as data grows and changes.

Range partitioning is just one of several partitioning strategies available in modern database systems, each suited for different scenarios and data characteristics. for specific database

LIST PARTITIONING:

List partitioning is another database partitioning technique where data is divided into partitions based on specific values in a chosen column. Each partition contains rows that match a predefined list of values. Here's a simple example of list partitioning in a table:

Scenario: Product Inventory Partitioning

Consider a product inventory database with a table named "Products." You want to partition the "Products" table based on product categories to improve query performance.

Original Table: Products

ProductID	ProductName	Category	Quantity
1	Laptop	Electronics	50
2	Chair	Furniture	30
3	Smartphone	Electronics	100
...	...	...	...

List Partitioning:

You decide to partition the "Products" table based on the "Category" column. You create partitions for specific categories:

- ➔Partition 1: Electronics
- ➔Partition 2: Furniture
- ➔Partition 3: Other (for categories not explicitly listed)

After partitioning, the data would be physically organized like this:

Partition 1: Electronics

ProductID	ProductName	Category	Quantity
1	Laptop	Electronics	50
3	Smartphone	Electronics	100
...	...	...	...

Partition 2: Furniture

ProductID	ProductName	Category	Quantity
2	Chair	Furniture	30
...	...	...	...

Partition 3: Other

ProductID	ProductName	Category	Quantity
...	...	...	...

Benefits of List Partitioning:

- **Simplified Data Organization:** List partitioning allows you to group similar data together based on predefined lists of values.
- **Enhanced Query Performance:** By partitioning on categories, queries that target specific categories can perform better since they only access the relevant partitions.

Considerations:

- Choose partitioning columns that align with your data access patterns.
- Keep the number of partitions manageable to avoid excessive complexity.

List partitioning is a useful strategy when you want to segregate data based on specific categories, types, or other predefined values. Like other partitioning techniques, it can significantly improve database performance and manageability for large datasets.

HASH PARTITIONING:

Hash partitioning is a database partitioning technique where data is distributed across partitions based on a hash function applied to a chosen column's values. This technique is useful for distributing data uniformly and balancing the load across partitions. Here's a simple example of hash partitioning in a table:

Scenario: User Data Hash Partitioning

Consider a user database with a table named "Users." You want to partition the "Users" table based on the user IDs to ensure even distribution of data.

Original Table: Users

UserID	Username	Email	Age
1	Abel	Abel@example.com	25
2	Bill	Bill@example.com	30
3	Catherine	Catherine@example.com	28
...	...	...	...

Hash Partitioning:

You decide to hash partition the "Users" table based on the "UserID" column. Let's say you choose to create 4 partitions.

After hash partitioning, the data would be distributed across partitions based on the hash values of the "UserID" column. For simplicity, let's assume the hash function returns values from 1 to 4.

Partition 1:

UserID	Username	Email	Age
1	Abel	Abel@example.com	25
5	Angus	Angus@example.com	22
...	...	...	...

Partition 2:

UserID	Username	Email	Age
2	Bill	Bill@example.com	30
6	Gertrude	Gertrude@example.com	29
...	...	...	...

Partition 3:

UserID	Username	Email	Age
3	Catherine	Catherine@example.com	28
7	Alphonsus	Alphonsus@example.com	26
...	...	...	...

Partition 4:

UserID	Username	Email	Age
4	Bart	Bart@example.com	23
8	Fritz	Fritz@example.com	24
...	...	...	...

Benefits of Hash Partitioning:

- **Uniform Data Distribution:** Hash partitioning ensures even distribution of data across partitions, preventing hotspots.

- **Predictable Query Performance:** Hash partitioning helps distribute the query load more evenly, resulting in predictable performance.

Considerations:

- Choose an appropriate number of partitions to match the expected data distribution.

- Hash partitioning works best when there's a relatively uniform distribution of data across the partitioning column.

Hash partitioning is particularly useful for large datasets where uniform distribution and balanced query performance are essential. It's often chosen when the data doesn't have a clear range or list-based partitioning criterion.

SHARDING:

Sharding, or vertical partitioning, involves distributing data and workload across multiple database instances or servers, known as shards. Each shard holds a portion of the data, typically based on a defined partitioning key. Sharding offers several advantages:

1. Scalability: Sharding enables horizontal scaling by distributing data across multiple shards. Each shard operates independently, allowing the system to handle increased data volumes and user concurrency effectively.

2. Performance: Sharding improves performance by reducing the data footprint on individual shards. Queries are directed to specific shards, minimizing data access and improving query response times.

3. Fault Isolation: Sharding enhances fault isolation, as a failure in one shard does not affect the entire system. Each shard operates independently, reducing the impact of failures and improving system availability.

SHARDING STRATEGIES:

RANGE-BASED SHARDING:

Data is divided based on a defined range of values, similar to range partitioning. Each shard holds data falling within specific ranges.

HASH-BASED SHARDING:

Data is distributed across shards using a hashing algorithm applied to a specific shard key. Hash-based sharding ensures even data distribution, but it may not support efficient range-based queries.

CONSIDERATIONS FOR IMPLEMENTATION:

1. Data Distribution: When partitioning or sharding, consider the distribution of data to ensure balanced loads across nodes or shards. Uneven data distribution can lead to performance bottlenecks and skewed query performance.

2. Partition/Shard Key Selection: Choosing an appropriate partition or shard key is crucial. It should distribute data evenly and align with query patterns to optimize query performance.

3. Query Routing: Implement an effective mechanism to route queries to the appropriate partitions or shards based on the partition/shard key. This ensures efficient data retrieval and maximizes the benefits of partitioning or sharding.

4. Data Consistency and Joins: Consider the impact of data consistency and join operations in partitioning or sharding scenarios. Joining data across partitions or shards can be challenging and may require additional considerations and techniques, such as distributed joins or denormalization.

5. Data Integrity and Recovery: Implement mechanisms to ensure data integrity and consistency across partitions or shards. This may involve distributed transaction management, replication, or backup and recovery strategies that account for distributed data.

6. Monitoring and Maintenance: Establish monitoring and maintenance practices to track the performance and health of partitions or shards. This includes monitoring resource utilization, data distribution, and addressing maintenance tasks like data rebalancing or shard/node failures.

7. Application Changes: Partitioning or sharding may require changes to the application layer to support distributed data access and query routing. Ensure that the application is designed to handle partitioned or sharded data effectively.

Partitioning and sharding are powerful techniques for optimizing scalability and performance in modern database systems. Partitioning involves horizontally dividing a database into smaller subsets, while sharding distributes data vertically across multiple shards or database instances. Both approaches offer benefits in terms of scalability, performance, and manageability. Careful consideration of partitioning or sharding strategies, such as range-based or hash-based approaches, along with key selection, data distribution, and query routing mechanisms, is essential for successful implementation. Additionally, addressing challenges related to data consistency, joins, data integrity, and monitoring ensures the effectiveness and reliability of partitioning or sharding solutions. By adopting these approaches and considering the specific requirements of the application, organizations can harness the power of partitioning and sharding to handle large-scale data processing, improve performance, and achieve scalability in their database systems.

CHAPTER 4. QUERY OPTIMIZATION TECHNIQUES

Query optimization is a central aspect of database performance tuning, aimed at improving the efficiency of query execution and overall system performance. Using techniques for efficient queries is a good starting point. In this chapter we learn efficient SQL anti-patterns, execution plans, and query tuning and optimization.

Let's start by showing a picture of the database query process:

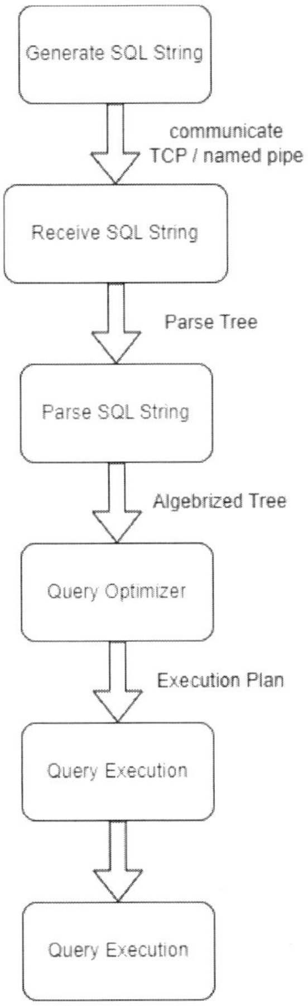

An efficient query takes into account the process steps and how each step can execute efficiently. During parsing, the parser will generate a representation of the query and pass that to the optimizer. The optimizer defines an algorithm to optimally run the SQL statement. However, one can help this process by using the following performance tips:

- Use appropriate indexes to speed up queries that filter or sort large datasets. Indexes are data structures that store a subset of the columns of a table, along with pointers to the corresponding rows. They can help the database engine find the relevant rows faster, without scanning the entire table. You can create indexes on columns that are frequently used in the WHERE or ORDER BY clauses of your queries.

- Avoid using functions on columns in the WHERE clause, as these can prevent the use of indexes. For example, instead of writing `WHERE UPPER(name) = 'JOHN'`, write `WHERE name = 'John'`. If required to perform case-insensitive comparisons, you can use the COLLATE operator.

- Use EXPLAIN to understand the execution plan of your query and identify potential bottlenecks. EXPLAIN is a keyword that you can prepend to any SQL query to get a detailed description of how the query will be executed by the database engine. It can show you information such as the tables and indexes involved, the order of operations, the estimated cost and time, and the number of rows returned or affected.

- Use SELECT * only when you need all the columns from a table. Otherwise, specify only the columns that you actually need in your query. This can reduce the amount of data that needs to be transferred from the database to your application, and also avoid unnecessary computations on columns that are not relevant.

- Use LIMIT and OFFSET for paging large results sets. If required to display only a subset of rows from a query, you can use the LIMIT clause to specify how many rows you want, and the OFFSET clause to specify how many rows you want to skip. For example, `SELECT * FROM products LIMIT 10 OFFSET 20` will return 10 rows starting from the 21st row. This can improve the performance of your query by avoiding fetching and processing more rows than needed.

- Use subqueries carefully and sparingly. Subqueries are queries that are nested inside another query, usually in the WHERE or SELECT clauses. They can be useful for performing complex calculations or filtering, but they can also introduce overhead and complexity to your query. Sometimes, you can rewrite a subquery using a JOIN or an EXISTS clause, which can be more efficient.

- Optimize your WHERE clause by using appropriate operators and conditions. The WHERE clause is one of the most important parts of a SQL query, as it determines which rows from the table will be returned or affected by the query. You should use operators and conditions that are compatible with the data type of the column, and avoid unnecessary or redundant conditions. For example, instead of writing `WHERE age >= 18 AND age < 65`, write `WHERE age BETWEEN 18 AND 64`.

- Use CASE instead of UPDATE when possible. The UPDATE statement modifies existing rows in a table, which can be expensive and time-consuming, especially if it involves large amounts of data or triggers. The CASE statement allows you to perform conditional logic on a column without modifying it. For example, instead of writing `UPDATE products SET price = price * 1.1 WHERE category = 'Electronics'`, write `SELECT name, CASE WHEN category = 'Electronics' THEN price * 1.1 ELSE price END AS price FROM products`.

One may consider skipping steps to increase performance. Skipping the SQL parsing step, including parameter binding, is not typically done to optimize database performance. SQL parsing is an essential part of query execution, and parameter binding is a crucial security measure to prevent SQL injection attacks and optimize query plan reuse.

Here's why SQL parsing and parameter binding are important:

1. **Query Parsing**: Parsing involves interpreting the SQL statement, identifying the tables involved, determining the join conditions, and creating a query plan. Parsing ensures that the database understands what the query is asking for and how to retrieve the data efficiently.

2. **Parameter Binding**: Parameter binding is the process of substituting parameters in a query with their actual values. This is important for security because it prevents malicious users from injecting harmful SQL code. Parameterized queries also help optimize query plan reuse, as the database can cache and reuse query plans with different parameter values.

3. **Query Optimization**:After parsing and parameter binding, the database's query optimizer determines the most efficient way to execute the query. This may involve selecting appropriate indexes, join methods, and access paths. Skipping these steps would result in suboptimal query execution.

While it might be tempting to skip parsing and binding to save time, doing so would compromise the integrity, security, and performance of your database queries. Instead, it's better to focus on optimizing these steps, such as using prepared statements and bind variables, to ensure efficient and secure query execution.

Considering other optimization techniques such as index creation, query rewriting, and database schema design improvements, rather than bypassing critical query processing steps will usually yield better results.

Query execution plans play a pivotal role in this optimization process. They provide insights into how the database engine processes queries and guides the selection of the most efficient execution strategy. This section explores the concept of query execution plans and their interpretation, highlighting their significance in optimizing query performance.

A query optimizer is an instrumental component of a database management system (DBMS) that's responsible for analyzing SQL queries and determining the most efficient way to retrieve the requested data. Many relational database management systems include a query optimizer. Here's an example:

Example: SQL Query Optimization

Consider a simplified "Employees" table with columns for "EmployeeID," "FirstName," "LastName," "Department," "Salary," and "HireDate." You want to retrieve the names of employees in the "Sales" department whose salaries are greater than $450,000.

```
SELECT FirstName, LastName
FROM Employees
WHERE Department = 'Sales' AND Salary > 450000;
```

When you submit this query to the DBMS, the query optimizer kicks in to determine the optimal execution plan. It considers various factors like available indexes, table statistics, and the complexity of the query.

Here's a simplified view of what the optimizer might do:

1. **Parse the Query:** The optimizer parses the SQL query to understand its structure and requirements.

2. **Access Path Selection:** It evaluates different access paths, such as full table scans or index scans, and chooses the most efficient one based on available statistics and indexes.

3. **Join Order and Method:** If there are multiple tables involved in the query, it determines the order in which to join them and the join method (e.g., nested loops, hash joins) for optimal performance.

4. **Filter Order:** It determines the order in which filters and conditions should be applied, optimizing the query for speed.

5. **Index Usage:** If there are suitable indexes, the optimizer decides whether to use them and how to access the data efficiently.

6. **Cost-Based Optimization:** It calculates the cost of different execution plans and selects the one with the lowest cost.

7. **Execution Plan Generation:** Finally, the optimizer generates an execution plan that details how the query will be executed, including which tables will be accessed, which indexes will be used, and the order of operations.

The goal of the query optimizer is to minimize the time and resources required to retrieve the requested data while ensuring data consistency and correctness. In the case, this optimization process is often automatic, but there are also mechanisms for database administrators to influence the optimizer's decisions, such as using hints to suggest specific optimization strategies.

UNDERSTANDING QUERY EXECUTION PLANS:

A query execution plan outlines the step-by-step process the database engine follows to execute a query. It represents the chosen query execution strategy, which includes the order of operations, access methods, join algorithms, and data retrieval techniques. The query execution plan is generated by the database optimizer, which analyzes the query and evaluates various execution strategies to identify the most optimal one. A query optimizer facilitates developing an optimal performing execution strategy. This robust query optimizer is central to efficiently develop a high-performing query execution plan.

INTERPRETING QUERY EXECUTION PLANS:

Interpreting query execution plans involves understanding the components and operators present in the plan and their implications on query performance. Here are key aspects to consider when interpreting query execution plans:

1. **Access Methods**: Access methods describe how the database engine retrieves data from tables or indexes. Common access methods include table scans, index scans, or index seeks. A table scan implies reading the entire table, while index scans or seeks indicate the use of indexes to locate specific data. Evaluating the access methods helps identify potential performance bottlenecks and determine the effectiveness of index usage.

2. **Join Operations**: combine data from multiple tables based on specified join conditions. Join algorithms, such as nested loop joins, hash joins, or merge joins, determine how the join is performed. Understanding the chosen join algorithm aids in assessing the complexity and efficiency of join operations. It also helps identify opportunities for optimizing join performance by ensuring appropriate indexes, join order, or join type selection.

3. **Filter and Predicate Evaluation**: define the conditions used to filter or limit the result set. The query execution plan reveals how these conditions are evaluated. Examining the placement and order of filters in the plan helps identify potential inefficiencies, such as unnecessary or non-sargable predicates that could impact query performance. A query predicate where the DBMS engine cannot take advantage of an index to speed the query is considered non-sargable. The term is derived from a contraction of Search ARGument ABLE. Conversely, the condition (or predicate) in a query is said to be sargable if the DBMS engine can take advantage of an index to speed up the execution of the query.

4. **Sort and Grouping Operations**: involve ordering and aggregating data based on specified criteria. The query execution plan provides insights into the presence and type of sort or grouping operators. Understanding these operations helps evaluate the impact on query performance, especially when large result sets or complex aggregations are involved.

5. **Data Retrieval Techniques**: The query execution plan indicates how data is retrieved, whether from disk, memory, or a combination of both. Assessing the data retrieval techniques helps identify potential I/O bottlenecks and assess the efficiency of data access.

OPTIMIZING QUERY PERFORMANCE USING EXECUTION PLANS:

Query execution plans serve as a valuable tool for query optimization. By interpreting and analyzing these plans, one can identify areas for optimization and implement the following strategies:

1. **Index Optimization**: Evaluate index usage and identify opportunities for creating, modifying, or dropping indexes to enhance data access efficiency. Ensure indexes align with the query execution plan and query patterns.
2. **Statistics Updates**: Query execution plans rely on accurate statistics about data distribution and cardinality. Regularly update statistics to ensure the optimizer has up-to-date information for making effective query execution decisions.
3. **Query Rewriting**: Analyze the query execution plan to identify query patterns that can be rewritten or optimized for better performance. Consider rewriting complex queries, removing redundant calculations, or restructuring join conditions based on the insights gained from the plan.
4. **Configuration and Resource Tuning**: The query execution plan may reveal performance bottlenecks related to resource usage. Use the plan to optimize database configurations, such as memory allocation, parallelism settings, or disk I/O configurations, to better align with the query execution requirements.
5. **Indexing Strategies and Materialized Views**: The query execution plan can guide the identification of potential index or materialized view opportunities to precompute and optimize specific query results.

Query execution plans are indispensable tools for understanding and optimizing query performance. By interpreting these plans, database administrators and developers gain insights into the execution strategies chosen by the optimizer, enabling them to identify optimization opportunities, evaluate access methods, join operations, filters, and sorting techniques. With this knowledge, they can make informed decisions regarding index optimization, query rewriting, configuration tuning, and other techniques to enhance query performance. By leveraging query execution plans effectively, organizations can achieve substantial improvements in database performance, enabling faster and more efficient query execution and improving overall system efficiency.

IDENTIFYING AND RESOLVING INEFFICIENT QUERIES

Inefficient queries can significantly impact the performance of database systems, leading to slow response times, increased resource utilization, and degraded user experience. Identifying and resolving inefficient queries is necessary and fundamental for optimizing database performance. This section discusses strategies for identifying inefficient queries and provides guidance on resolving them to improve overall system efficiency.

IDENTIFYING INEFFICIENT QUERIES:

Identifying inefficient queries involves monitoring and analyzing query performance to pinpoint the queries causing performance bottlenecks. Here are key strategies to identify inefficient queries:

1. **Query Profiling and Monitoring**: Implement query profiling and monitoring tools to capture query execution statistics, such as execution time, CPU usage, disk I/O, and memory consumption. Analyze these metrics to identify queries with long execution times or excessive resource consumption.

2. **Database Performance Metrics**: Leverage database performance monitoring tools to track system-level metrics, such as CPU utilization, memory usage, disk I/O, and network throughput. High resource utilization during query execution can indicate inefficient queries.

3. **Query Execution Plans**: Examine query execution plans, which outline the steps taken by the database optimizer to execute queries. Look for access methods, join operations, and sorting techniques that may indicate sub-optimal query execution strategies. Identify full table scans, missing or ineffective indexes, and excessive sorting or grouping operations.

4. **Query Log Analysis**: Analyze query logs or statement logs to identify queries with high execution counts or prolonged execution times. Look for recurring patterns or frequently executed queries that may be candidates for optimization.

RESOLVING INEFFICIENT QUERIES:

Once inefficient queries are identified, several strategies can be employed to resolve them and improve performance:

1. **Index Optimization**: Evaluate index usage and consider creating or modifying indexes to align with query execution plans. Ensure that indexes cover frequently accessed columns and are not redundant. Remove unused or ineffective indexes that consume resources without providing benefits.

2. **Query Rewriting**: Rewrite inefficient queries to optimize their performance. Consider eliminating unnecessary joins, reducing the number of columns fetched, simplifying complex expressions or subqueries, and restructuring conditions or

filters. Rewrite queries to leverage index usage and exploit the strengths of the database optimizer.

3. **Schema Optimization**: Review the database schema to ensure it is properly designed and normalized. Evaluate table relationships, foreign keys, and constraints to minimize redundant or excessive data retrieval. Optimize the database schema based on the specific requirements of the application and the queries executed.

4. **Data Caching and Materialized Views**: Implement data caching mechanisms or materialized views to precompute and store frequently accessed query results. This reduces the need for repeated expensive computations and improves query response time.

5. **Parameterization and Prepared Statements**: Use parameterized queries and prepared statements to promote query plan reuse and minimize the overhead associated with query compilation and optimization.

6. **Database Configuration Tuning**: Optimize database configuration parameters, such as memory allocation, parallelism settings, disk I/O settings, and cache sizes, to better align with the workload and query patterns. Adjusting these configurations can have a significant impact on query performance.

7. **Regular Maintenance**: Perform regular maintenance tasks, such as updating statistics, rebuilding indexes, and reorganizing database objects. These tasks ensure that query optimizers have accurate information to make effective execution decisions.

8. **Query Performance Testing**: Conduct performance testing and benchmarking to measure the impact of query optimization strategies. Validate the effectiveness of optimizations before deploying them to production environments.

Identifying and resolving inefficient queries is commonplace and necessary for optimizing database performance and improving overall system efficiency. By leveraging query profiling, performance monitoring, execution plans, and log analysis, organizations can identify queries causing performance bottlenecks. Applying strategies such as index optimization, query rewriting, schema optimization, data caching, and configuration tuning, along with regular maintenance, helps address the root causes of inefficiency. Continuously monitoring and fine-tuning the database system based on query performance testing enables ongoing optimization and ensures long-term efficiency gains. By proactively addressing inefficient queries, organizations can achieve faster query execution, improved system performance, and enhanced user satisfaction.

JOIN OPTIMIZATION AND QUERY REWRITING: ENHANCING QUERY PERFORMANCE

Join operations play a significant and ubiquitous role in relational databases, but they can often be a source of performance bottlenecks. Join optimization techniques and query rewriting strategies offer effective approaches to improve query performance by optimizing join operations. This section explores join optimization and query rewriting techniques, highlighting their significance in enhancing query performance.

JOIN OPTIMIZATION:

Join optimization focuses on improving the efficiency of join operations by selecting the most optimal join algorithms, join orders, and access methods. The objective is to minimize the computational complexity and reduce the overall execution time of join queries. Join optimization techniques include join algorithms, join order optimization, and join predicate pushdown.

JOIN ALGORITHMS:

Database systems employ various join algorithms, such as nested loop joins, hash joins, and merge joins. Each algorithm has its strengths and weaknesses, depending on factors such as data size, data distribution, and available memory. Optimizers analyze query statistics and characteristics to determine the most efficient join algorithm for a given query.

Let's consider an example of using a join algorithm as a join optimization technique. We'll use a simplified scenario involving two tables: "Orders" and "Clients" in a hypothetical database.

Scenario: Join Optimization with Hash Join Algorithm

Suppose you have two tables: "Orders" and "Clients," and you want to retrieve a list of orders with client information. You're considering using a hash join algorithm to optimize this query.

Tables:

Orders Table:

OrderID	OrderDate	ClientID	TotalAmount
1	2024-01-01	101	1000
2	2024-01-05	102	1500
3	2024-01-10	101	800
...	...	...	...

Clients Table:

ClientID	FirstName	LastName	Email
101	Abel	Dickens	Bill@example.com
102	Bill	Hendricks	
...	...	...	...

Query:

Obtaining a list of orders with client information (joining on the "ClientID" column):

```
SELECT Orders.OrderID, Orders.OrderDate,
Clients.FirstName, Clients.LastName
FROM Orders
JOIN Clients ON Orders.ClientID = Clients.ClientID;
```

Using Hash Join Algorithm:

1. The hash join algorithm involves two phases: a build phase and a probe phase.
2. In the build phase, the database system reads the "Clients" table and builds a hash table based on the "ClientID" column.
3. In the probe phase, the database system scans the "Orders" table. For each row in the "Orders" table, it calculates the hash value of the "ClientID" and looks up the corresponding entry in the hash table.

4. When a match is found, the data from both tables is combined to produce the result.

Advantages of Hash Join Algorithm:

➢Hash join is particularly efficient for joining large tables, especially when there is no pre-existing index on the join columns.

➢It can be highly parallelized, making it suitable for multi-core processors and distributed database systems.

➢Hash joins typically perform well for equi-joins (joins using equality conditions).

Note: The database system's query optimizer will often choose the appropriate join algorithm (e.g., hash join, nested loop join, merge join) based on factors like table size, available indexes, and query complexity to optimize performance automatically.

In this example, we used a hash join algorithm to optimize the query by efficiently joining the "Orders" and "Clients" tables based on the "ClientID" column, resulting in a list of orders with client information.

JOIN ORDER OPTIMIZATION:

The order in which tables are joined can significantly impact query performance. Join order optimization involves evaluating different join orders and selecting the most efficient sequence of table joins. Dynamic programming algorithms, cost-based optimizers, or heuristics-based approaches are used to explore the search space and identify the optimal join order.

Join order optimization is a technique used to determine the most efficient order in which to join tables in a SQL query. Let's consider an example where join order optimization is applied to improve query performance. In this scenario, we have three tables: "Orders," "Clients," and "Products."

Orders Table:

OrderID	**OrderDate**	**ClientID**	**ProductID**	**Quantity**
1	2024-01-01	101	1	3
2	2024-01-05	102	2	2
3	2024-01-10	101	1	1
...	...	...	...	...

Clients Table:

ClientID	FirstName	LastName	Email
101	Abel	Dickens	Abel@example.com
102	Bill	Hendricks	Bill@example.com
...	...	...	...

Products Table:

ProductID	ProductName	Category
1	Laptop	Electronics
2	Chair	Furniture
...	...	...

Query Without Join Order Optimization:

Suppose we need a list of orders with client information and the product details for each order. Without join order optimization, the query might join the tables in the order they are mentioned in the query:

```
SELECT Orders.OrderID, Orders.OrderDate,
Clients.FirstName, Clients.LastName,
Products.ProductName

FROM Orders
JOIN Clients ON Orders.ClientID = Clients.ClientID
JOIN Products ON Orders.ProductID = Products.ProductID;
```

Query With Join Order Optimization:

With join order optimization, the database's query optimizer determines the most efficient join order. It considers factors such as table sizes, available indexes, and join conditions. Here's the optimized query:

```
SELECT Orders.OrderID, Orders.OrderDate,
Clients.FirstName, Clients.LastName,
Products.ProductName

FROM Clients
JOIN Orders ON Clients.ClientID = Orders.ClientID
JOIN Products ON Orders.ProductID = Products.ProductID;
```

Advantages of Join Order Optimization:

➢**Improved Query Performance:** By choosing the optimal join order, the database can reduce the amount of data processed and accessed during the query execution, resulting in faster query performance.

➢**Resource Efficiency:** Join order optimization can help minimize CPU and memory usage, making queries more resource-efficient.

➢**Consistent Results:** Join order optimization ensures that the query's results are still correct while optimizing the execution plan.

In this example, the database optimizer determined that joining the "Clients" table first before joining "Orders" and "Products" would result in a more efficient query execution plan. This demonstrates the importance of join order optimization in achieving better query performance.

Optimizing join order without the assistance of a database optimizer can be challenging, but you can follow some rules of thumb based on your knowledge of the data and query patterns. Keep in mind that these rules are general guidelines and might not always be optimal for every situation. Here are some tips:

1. **Start with Small Tables:** As a general rule, start by joining smaller tables first. This can reduce the overall dataset size early in the query execution.

2. **Consider Filter Predicates:** If you have filter predicates in your WHERE clause that significantly reduce the number of rows from a particular table, consider joining that table early. This can minimize the number of rows that need to be joined with other tables.

3. **Use Indexes:** If you have appropriate indexes on the join columns, consider joining tables that can utilize these indexes for efficient lookups first. This can speed up the join process.

4. **Think about Data Distribution:** Consider the distribution of data in your tables. If one table has a significantly skewed distribution (e.g., one value occurs much

more frequently than others), it might be better to join that table later to avoid creating a bottleneck.

5. **Avoid Cross Joins:** Avoid creating cross joins (Cartesian products) whenever possible. Cross joins generate a large number of rows and can be extremely slow. Ensure that you have appropriate join conditions in place.
6. **Use Subqueries:** In some cases, breaking down complex joins into multiple subqueries can improve performance. Subqueries can limit the number of rows being joined at once.
7. **Understand Data Relationships:** Consider the logical relationships between tables. Tables with one-to-many or many-to-one relationships should be joined accordingly. Starting with the "one" side of the relationship is often a good choice.
8. **Analyze Execution Plans:** If you're working with a relational database system that provides query execution plans (e.g., EXPLAIN PLAN), analyze the plans for different join orders to identify potential bottlenecks.
9. **Test and Benchmark:** Test different join orders and benchmark the query performance to determine which order works best for your specific data and query workload.
10. **Consider Query Complexity:** Think about the overall query complexity. More complex queries might benefit from a different join order than simpler queries.

Remember that the optimal join order can vary significantly depending on the specifics of your data, indexes, and the query itself. Experimentation and performance testing are essential to fine-tune join order optimization for your particular use case.

JOIN PREDICATE PUSHDOWN:

Predicate (condition) pushdown involves pushing join predicates to the earliest possible stage in query processing. By applying predicates early, unnecessary rows are filtered out before the join operation, reducing the amount of data that needs to be processed. This optimization technique can lead to significant performance improvements, especially when joining large tables.

Join predicate pushdown is an optimization technique that involves pushing join conditions (predicates) closer to the source tables in a query to minimize the amount of data processed during the join. Let's consider an example where join predicate pushdown is applied to improve query performance.

Scenario: Join Predicate Pushdown Optimization

Suppose you have two tables: "Orders" and "Clients" in a relational database. You want to retrieve a list of orders made by clients with a specific email domain (e.g.,

"@example.com"). In this example, you'll apply join predicate pushdown to minimize data transfer and processing.

Orders Table:

OrderID	ClientID	OrderDate	TotalAmount
1	101	2024-01-01	1000
2	102	2024-01-05	1500
3	103	2024-01-10	800
...	...	...	...

Clients Table:

ClientID	FirstName	LastName	Email
101	Abel	Dickens	Abel@example.com
102	Bill	Hendricks	Bill@example.com
103	Kate	Newton	Kate@gmail.com
...	...	...	...

Query Without Join Predicate Pushdown:

Without join predicate pushdown, the query might initially retrieve all rows from the "Clients" table before filtering by the email domain:

```
SELECT Orders.OrderID, Orders.OrderDate, Clients.FirstName,
Clients.LastName
FROM Orders
JOIN Clients ON Orders.ClientID = Clients.ClientID
WHERE Clients.Email LIKE '%@example.com';
```

Query With Join Predicate Pushdown:

With join predicate pushdown, you can push the email domain filter closer to the "Clients" table, so only relevant rows are joined with the "Orders" table:

```
SELECT Orders.OrderID, Orders.OrderDate, Clients.FirstName,
Clients.LastName

FROM Orders
JOIN (
    SELECT ClientID, FirstName, LastName
    FROM Clients
    WHERE Email LIKE '%@example.com'
) AS Clients
ON Orders.ClientID = Clients.ClientID;
```

Advantages of Join Predicate Pushdown:

- **Reduced Data Transfer:** By filtering the "Clients" table before the join, you minimize the amount of data transferred between tables.
- **Improved Query Performance:** The database engine processes fewer rows during the join, leading to faster query execution.
- **Efficient Resource Usage:** Join predicate pushdown optimizes CPU and memory usage, especially when dealing with large tables.

In this example, join predicate pushdown is applied by first filtering the "Clients" table to only include rows with the specified email domain. Then, the join is performed with a smaller, filtered dataset, resulting in improved query performance.

QUERY REWRITING:

Query rewriting involves transforming queries to equivalent forms that optimize performance. By modifying the query structure or introducing new constructs, query rewriting aims to simplify expressions, reduce unnecessary calculations, and improve query execution efficiency. Common query rewriting techniques include:

SUBQUERY UNFOLDING:

Subquery unfolding, also known as subquery materialization or subquery unnesting, is a query rewriting technique where a subquery is replaced with its result, essentially transforming it into a join or a derived table. This involves replacing subqueries with join operations or derived tables. This technique transforms complex subqueries into more efficient forms that leverage joins and eliminate unnecessary nesting. By reducing the number of subqueries, query performance can be significantly improved.

Let's consider an example where subquery unfolding is applied to improve query performance.

Scenario: Subquery Unfolding Optimization

Suppose you have two tables: "Orders" and "Clients" in a relational database. You want to retrieve a list of clients who have placed orders with a total amount greater than $1,000. In this example, you'll apply subquery unfolding to optimize the query.

Orders Table:

OrderID	ClientID	OrderDate	TotalAmount
1	101	2024-01-01	1000
2	102	2024-01-05	1500
3	101	2024-01-10	800
...	...	...	...

Clients Table:

ClientID	FirstName	LastName	Email
101	Abel	Dickens	Abel@example.com
102	Bill	Hendricks	Bill@example.com
103	Kate	Newton	Kate@example.com
...	...	...	...

Query Without Subquery Unfolding:

Without subquery unfolding, you might use a subquery to calculate the total amount for each client and then filter the clients who meet the criteria:

```
SELECT FirstName, LastName
FROM Clients
WHERE ClientID IN (
   SELECT ClientID
   FROM Orders
   GROUP BY ClientID
   HAVING SUM(TotalAmount) > 1000
);
```

Query With Subquery Unfolding:

With subquery unfolding, you can calculate the total amount for each client using a subquery and then join the result with the "Clients" table:

```
SELECT C.FirstName, C.LastName
FROM Clients AS C
JOIN (
   SELECT ClientID
   FROM Orders
   GROUP BY ClientID
   HAVING SUM(TotalAmount) > 1000
) AS TotalAmounts
ON C.ClientID = TotalAmounts.ClientID;
```

Advantages of Subquery Unfolding:

➢**Improved Query Performance:** Subquery unfolding can lead to better query execution plans, especially when dealing with large datasets, as it can optimize the query's logical processing order.

➢**Simplified Query:** The query becomes more straightforward and easier to understand since it doesn't involve subqueries.

➢**Optimized Resource Usage:** Subquery unfolding can help optimize CPU and memory usage by reducing redundant calculations.

In this example, subquery unfolding is applied by first calculating the total amount for each client using a subquery and then joining the result with the "Clients" table to retrieve the desired client information. This optimization can improve query performance and readability.

PREDICATE SIMPLIFICATION:

Predicate simplification is a query optimization technique that involves simplifying complex conditions in a query to make it more efficient and readable. Let's consider an example where predicate simplification is applied to improve a query. It involves simplifying complex predicates by applying logical transformations or eliminating redundant conditions. This optimization technique reduces the number of conditions evaluated during query execution, improving performance. Techniques like Boolean algebra rules, De Morgan's laws, and equivalence transformations are used to simplify predicates.

Scenario: Predicate Simplification Optimization

Suppose you have a database table named "Employees" with information about employees, including their ages. You want to retrieve a list of employees who are older than 25 years old but not older than 30 years old. In this example, you'll apply predicate simplification to optimize the query.

Employees Table:

EmployeeID	FirstName	LastName	Age
1	Abel	Dickens	28
2	Bill	Hendricks	30
3	Kate	Newton	24
...	...	...	...

Query Without Predicate Simplification:

Without predicate simplification, you might express the query with two separate conditions joined by the logical AND operator:

```
SELECT EmployeeID, FirstName, LastName, Age
FROM Employees
WHERE Age > 25 AND Age <= 30;
```

Query With Predicate Simplification:

With predicate simplification, you can express the same query more succinctly using a single range condition:

```
SELECT EmployeeID, FirstName, LastName, Age
FROM Employees
WHERE Age BETWEEN 26 AND 30;
```

Advantages of Predicate Simplification:

- **Improved Query Readability:** Predicate simplification makes the query more concise and easier to understand.
- **Potential Query Optimization:** Some database systems may optimize range conditions more effectively than separate conditions, potentially leading to better query performance.
- **Reduced Risk of Error:** Simplified predicates are less prone to logical errors or typos compared to complex conditions.

In this example, predicate simplification is applied by consolidating two separate age conditions into a single range condition, which makes the query more readable and may potentially help in query optimization.

In the following example we use De Morgan's laws for logical simplification:
De Morgan's laws are a set of principles in Boolean algebra that describe the relationships between logical operations like AND, OR, and NOT. They can be applied in SQL queries to simplify complex conditions. Let's consider an example where De Morgan's laws are used to simplify a query condition.

Scenario: De Morgan's Laws Optimization

Suppose you have a database table named "Products" with information about various products, including their prices and quantities. You want to retrieve a list of products that are either priced above $100 or have a quantity in stock less than 10. You'll apply De Morgan's laws to simplify the query condition.

Products Table:

ProductID	ProductName	Price	QuantityInStock
1	Laptop	800	5
2	Smartphone	400	15
3	Headphones	50	25
...	...	...	...

Query Without De Morgan's Laws:

Without applying De Morgan's laws, you might express the query condition using the logical OR operator:

```
SELECT ProductID, ProductName, Price, QuantityInStock
FROM Products
WHERE Price > 100 OR QuantityInStock < 10;
```

Query With De Morgan's Laws:

Using De Morgan's laws, you can express the same query condition more succinctly by applying the logical NOT operator and the AND operator:

```
SELECT ProductID, ProductName, Price, QuantityInStock
FROM Products
WHERE NOT (Price <= 100 AND QuantityInStock >= 10);
```

Advantages of Using De Morgan's Laws:

- **Simplified Query:** De Morgan's laws help simplify complex conditions, making the query more readable.
- **Improved Logical Clarity:** The use of De Morgan's laws can make the logic of the condition more explicit and easier to understand.
- **Potential Query Optimization:** In some cases, database systems may optimize queries with simplified conditions more effectively.

In this example, De Morgan's laws are applied by simplifying the condition that combines price and quantity criteria using the logical NOT operator and the AND operator. This results in a more concise and logically equivalent query condition.

AGGREGATE PUSHDOWN:

Aggregate pushdown is an optimization technique used in database query processing to improve the performance of aggregate queries. This involves pushing aggregate operations closer to the data source to minimize the amount of data processed. By aggregating data early in the query execution process, unnecessary rows can be eliminated, leading to improved query performance. Aggregate pushdown is particularly useful when dealing with large datasets and complex aggregation operations. This technique involves pushing the aggregation operations closer to the data source, often within subqueries or derived tables, to minimize the amount of data that needs to be processed and transferred. This can significantly enhance query performance, especially when dealing with large datasets.

Here's an example to illustrate aggregate pushdown:

Scenario: Aggregate Pushdown Optimization

Suppose you have a database containing a "Sales" table that records sales transactions, including the sale amount, date, and the salesperson responsible. You want to retrieve the total sales amount for each salesperson. In this scenario, you'll apply aggregate pushdown to optimize the query.

Sales Table:

SaleID	Salesperson	SaleDate	SaleAmount
1	Abel	2024-01-01	500
2	Bill	2024-01-02	700
3	Abel	2024-01-03	400
...	...	...	...

Query Without Aggregate Pushdown:

Without aggregate pushdown, you might write a query that retrieves all sales data and performs the aggregation in the main query:

```
SELECT Salesperson, SUM(SaleAmount) AS TotalSales
FROM Sales
GROUP BY Salesperson;
```

Query With Aggregate Pushdown:

With aggregate pushdown, you can perform the aggregation within a subquery or derived table, and then join the result with the original table:

```
SELECT S.Salesperson, A.TotalSales
FROM (
   SELECT Salesperson, SUM(SaleAmount) AS TotalSales
   FROM Sales
   GROUP BY Salesperson
) AS A
JOIN Sales AS S ON A.Salesperson = S.Salesperson;
```

Advantages of Aggregate Pushdown:

- **Reduced Data Transfer:** Aggregate pushdown minimizes the amount of data transferred between tables, which is particularly beneficial when dealing with large datasets.
- **Improved Query Performance:** By performing aggregation on a smaller subset of data, the query execution time is reduced, resulting in faster query performance.
- **Optimized Resource Usage:** Aggregate pushdown can optimize CPU and memory usage, leading to more efficient query processing.

In this example, aggregate pushdown is applied by first aggregating the sales data within a subquery and then joining the aggregated result with the original "Sales" table. This optimization reduces the amount of data involved in the aggregation operation and can lead to significant performance improvements for aggregate queries.

BENEFITS AND CONSIDERATIONS:

Join optimization and query rewriting techniques offer several benefits for query performance optimization:

1. **Enhanced Query Performance**: By optimizing join operations and rewriting queries for efficiency, query execution times can be significantly reduced. This leads to faster response times, improved system throughput, and enhanced user experience.
2. **Reduced Resource Consumption**: Efficient join operations and optimized queries minimize resource utilization, such as CPU and memory, resulting in more efficient utilization of system resources. This allows the database system to handle higher user concurrency and scale effectively.

3. **Improved Scalability**: Optimized join operations and query structures contribute to improved scalability by reducing the computational complexity of queries. Scalability is vital for handling increasing data volumes and user loads.

While join optimization and query rewriting offer significant benefits, some considerations should be taken into account:

1. **Query Complexity**: The effectiveness of join optimization and query rewriting techniques depends on the complexity and structure of the queries. Highly complex queries may require a more in-depth analysis and advanced optimization strategies.

2. **Over-Optimization**: Over-optimization should be avoided, as it can lead to increased query compilation time and decreased maintainability. It is essential to strike a balance between optimization efforts and query complexity.

3. **Query Patterns**: Join optimization and query rewriting techniques should be applied based on query patterns and access patterns specific to the database and application. Not all queries may benefit equally from these techniques.

Join optimization and query rewriting techniques are essential for improving query performance and enhancing overall database system efficiency. By optimizing join algorithms, join orders, and query structures, query execution times can be significantly reduced. This leads to faster response times, reduced resource consumption, and improved scalability. Understanding the benefits and considerations associated with join optimization and query rewriting enables organizations to implement effective strategies that optimize query performance and deliver optimal results for their database systems.

CHAPTER 5. CONCURRENCY CONTROL AND LOCKING

This chapter explores concurrency control and locking and their relation to database performance in more detail.

CONCURRENCY CONTROL AND PERFORMANCE TUNING

Concurrency control is the mechanism that ensures multiple transactions can operate on a database simultaneously without violating data integrity. In a multi-user environment, concurrent access to data is crucial to maintain application responsiveness and user satisfaction. However, managing concurrent transactions introduces challenges like data conflicts, resource contention, and potential data anomalies. To strike a balance between maintaining data consistency and maximizing database performance, efficient concurrency control strategies are required. Concurrent access to data is crucial to maintain application responsiveness, especially in multi-user environments where multiple users or processes are interacting with a database simultaneously. This concurrency allows for efficient utilization of system resources and enables applications to provide timely responses to user requests. Several reasons illustrate why concurrent access to data is essential for maintaining application responsiveness:

1. **Minimized Resource Under-Utilization:** In a single-user environment, if one operation is being performed, other resources remain idle. Concurrent access ensures that when one operation is waiting for a resource (such as a database query), other operations can proceed, making better use of available system resources like CPU and memory.

2. **Improved User Experience:** Users expect applications to respond quickly, regardless of the number of users using the system concurrently. Concurrency allows multiple users to interact with the application simultaneously, ensuring that each user gets a responsive experience without excessive waiting times.

3. **Optimized Throughput:** Concurrent access allows a system to handle a larger number of requests in a given time period. Without concurrency, the system might become a bottleneck, leading to delays and potentially causing system downtime due to resource exhaustion.

4. **Enhanced Scalability:** Applications that support concurrent access can more easily scale to accommodate a growing user base. By efficiently handling multiple requests simultaneously, the application can accommodate increased demand without sacrificing responsiveness.

5. **Faster Transaction Processing:** Many applications involve transactions that require multiple steps to be completed together. Concurrency allows multiple transactions to proceed concurrently, minimizing the overall time required to process all transactions.

6. **Real-Time Data Updates:** In scenarios where data changes frequently, concurrent access allows users to access the most up-to-date information in real time. Without concurrency, users might have to wait for previous operations to complete before accessing the latest data.

7. **Multi-User Collaboration:** Collaboration applications, such as document editing platforms, benefit from concurrent access, enabling multiple users to edit and view documents simultaneously without interfering with each other's work.

8. **Batch Processing Efficiency:** In scenarios involving batch processing or data analysis, concurrent access allows multiple tasks to be executed simultaneously, reducing the time required to complete large-scale data operations.

9. **Improved System Utilization:** For distributed systems, concurrent access to resources across multiple nodes ensures better utilization of distributed computing resources, leading to higher efficiency and faster processing.

In modern computing environments, where users demand responsiveness and applications are often accessed by numerous users concurrently, the ability to handle multiple requests and tasks simultaneously is requisite. Concurrency control mechanisms ensure that multiple users can access and modify data concurrently while maintaining data integrity. Efficiently managing concurrent access allows applications to provide a seamless and responsive experience to users, even under heavy workloads, ultimately contributing to user satisfaction and business success.

TRANSACTION MANAGEMENT:

Efficient transaction management is essential for both data integrity and database performance. Transactions encapsulate multiple operations into a single unit of work, ensuring that either all operations are completed successfully or none are. ACID properties (Atomicity, Consistency, Isolation, Durability) guarantee transaction reliability, and choosing the appropriate isolation level is crucial for performance. Selecting higher isolation levels like Serializable might provide stronger data consistency but can lead to increased contention, impacting performance. Balancing transaction isolation with performance requirements is a fundamental consideration. Higher transaction isolation levels, such as Serializable, offer stronger data consistency by ensuring that transactions are executed in a way that preserves the illusion that they are executed one after the other, even when multiple transactions are executed concurrently. While this level of data consistency is crucial for maintaining accurate and reliable data, it can lead to increased contention and negatively impact performance due to its restrictive nature.

Here's how higher transaction isolation levels like Serializable provide stronger data consistency and the potential impact on performance:

- **Strong Data Consistency:** In a Serializable isolation level, transactions are executed in a way that avoids most anomalies, such as dirty reads, non-repeatable reads, and phantom reads. This level of consistency ensures that the results of transactions are as if they were executed in a serial order, providing a strong guarantee of data accuracy.

- **Locking and Contention:** To achieve strong data consistency, Serializable isolation often involves placing more restrictive locks on data. Locks prevent concurrent transactions from accessing or modifying the same data simultaneously. While this prevents anomalies, it can also result in increased contention for locked resources.

- **Performance Impact:** Contention occurs when multiple transactions are blocked or delayed due to locked resources. In a Serializable isolation level, the frequency of locks and contention is higher because transactions are required to acquire locks on a broader range of data. This can lead to transactions being delayed as they wait for other transactions to release their locks.

- **Reduced Concurrency:** Serializable isolation limits the degree of concurrency that can be achieved because transactions often need to wait for locks held by other transactions to be released. This reduction in concurrency can lead to lower throughput and increased response times for transactions, impacting the application's overall performance.

- **Deadlocks:** The increased use of locks in Serializable isolation raises the likelihood of deadlocks. Deadlocks occur when two or more transactions are waiting for each other to release locks, resulting in a standstill where no progress can be made. Deadlocks require detection and resolution mechanisms, further affecting performance.

- **Use Case Consideration:** While Serializable isolation is essential for scenarios that demand strict data consistency, such as financial transactions or critical data updates, it might not be necessary for all database operations. For read-heavy workloads or scenarios where occasional data anomalies are acceptable, lower isolation levels like Read Committed or Repeatable Read could provide a better balance between performance and consistency.

Higher transaction isolation levels like Serializable are necessary for maintaining strong data consistency in scenarios where data accuracy is paramount. However, the restrictive locking mechanisms associated with these isolation levels can result in increased contention, reduced concurrency, and potentially degraded performance. Careful consideration of the application's requirements, data access patterns, and the trade-offs between data consistency and performance is necessary to determine the appropriate isolation level to be used for each scenario.

CONCURRENCY CONTROL MECHANISMS:

Concurrency control mechanisms are techniques used in database management systems (DBMS) to manage the simultaneous execution of multiple transactions while maintaining data consistency and integrity. These mechanisms ensure that transactions can access and modify data in a controlled and coordinated manner, preventing conflicts that could lead to data anomalies. However, these mechanisms can also impact performance due to the additional overhead they introduce.

Here are common concurrency control mechanisms and their effects on performance:

1. LOCK-BASED CONCURRENCY CONTROL:

Lock-based concurrency control involves using locks to control access to data. Transactions request locks before accessing data and release them when done. There are two primary types of locks:

Shared Locks: Allow multiple transactions to read data concurrently but prevent write access.

Exclusive Locks: Allow a single transaction exclusive read and write access to data.

Effects on Performance:

Contention: While locks prevent conflicts, they can lead to contention when multiple transactions attempt to access the same data simultaneously. Contention can result in blocking, where one transaction waits for another to release a lock, leading to decreased throughput and longer response times.

Deadlocks: Poorly managed lock acquisition and release can result in deadlocks, where transactions are paused waiting for each other to release locks. Deadlocks require detection and resolution mechanisms, which introduce additional overhead and performance impact.

2. TIMESTAMP ORDERING CONCURRENCY CONTROL:

Timestamp ordering assigns a unique timestamp to each transaction. Transactions are ordered based on their timestamps, and conflicts are resolved by comparing timestamps. Transactions with older timestamps are rolled back or delayed if they conflict with newer transactions.

Effects on Performance:

Serialization: Timestamp ordering guarantees serializability, ensuring strong data consistency. However, the need to manage transaction timestamps and perform conflict resolution can introduce overhead that impacts performance.

Potential Delays: Older transactions might be delayed or rolled back if they conflict with newer transactions, potentially affecting response times and transaction completion rates.

3. OPTIMISTIC CONCURRENCY CONTROL:

Optimistic concurrency control assumes that conflicts are rare. Transactions are executed without acquiring locks, and before committing, a validation phase checks if conflicts have occurred. If no conflicts are detected, the transaction is committed; otherwise, it's rolled back and retried.

Effects on Performance:

Reduced Contention: Optimistic concurrency control reduces the need for locking, leading to fewer conflicts and less contention. This can improve throughput and response times.

Retry Overhead: The validation phase introduces an overhead of retrying transactions in case of conflicts. While this overhead is generally lower than constant locking, it can impact performance, especially with high contention.

Concurrency control mechanisms aim to strike a balance between ensuring data consistency and maintaining performance. The choice of mechanism depends on factors such as application requirements, workload patterns, and data access characteristics. While these mechanisms provide data integrity, they also introduce trade-offs, including potential performance impact due to locking, contention, deadlocks, or retry overhead. Performance tuning involves carefully selecting and configuring concurrency control mechanisms that align with the application's performance and consistency requirements while mitigating their potential downsides.

LOCKING PROTOCOLS:

Locking protocols are an integral part of concurrency control mechanisms in database management systems (DBMS). They regulate how transactions acquire and release locks on data items to ensure data consistency and integrity in a multi-user environment. While locking protocols are essential for maintaining data correctness, they can also have a significant impact on performance. Let's explore in detail how locking protocols affect performance:

1. CONTENTION AND BLOCKING:

Locking protocols, particularly those that use shared and exclusive locks, can lead to contention when multiple transactions compete for the same data item. Contention occurs when one transaction is blocked from proceeding due to the locks held by other transactions. High contention levels can result in transaction delays and decreased throughput, leading to slower response times and degraded application performance.

2. LOCK GRANULARITY:

Locking protocols offer different levels of granularity, such as row-level, page-level, or table-level locking. The choice of granularity impacts both performance and concurrency. Finer-grained locks (e.g., row-level) provide higher concurrency by allowing transactions to access different parts of the data simultaneously. However, they can also lead to increased lock management overhead, potentially impacting performance.

3. DEADLOCKS:

Locking protocols, if not carefully managed, can lead to deadlocks. Deadlocks occur when two or more transactions are paused waiting for each other to release locks, causing a standstill. DBMSs usually detect deadlocks and initiate mechanisms to resolve them. However, the detection and resolution process adds overhead and can temporarily impact performance.

4. LOCK ESCALATION:

Some locking protocols involve lock escalation, where multiple fine-grained locks are converted into fewer coarser-grained locks to reduce overhead. While lock escalation can optimize lock management, it can also lead to increased contention and contention-related performance issues.

5. LOCK TIMEOUT AND DEADLOCK DETECTION:

Locking protocols often include mechanisms to handle blocked transactions, such as lock timeouts and deadlock detection. These mechanisms introduce overhead, as the DBMS needs to constantly monitor and manage blocked transactions. Adjusting timeout values to be too aggressive can lead to unnecessary transaction rollbacks, affecting performance.

6. REDUCED CONCURRENCY:

While locking protocols provide data consistency, they inherently limit concurrency. Transactions might be blocked from accessing data due to locks held by other transactions, resulting in lower throughput and potentially increased response times. The more restrictive the locking protocol, the higher the potential impact on concurrency and performance.

7. LOCKING OVERHEAD:

Locking protocols introduce additional overhead, such as acquiring and releasing locks, maintaining a lock table, and resolving lock conflicts. This overhead impacts the execution time of transactions, potentially leading to longer response times and lower throughput.

8. LOCK MANAGEMENT:

Managing locks requires coordination and memory resources. The more locks that need to be tracked and managed, the more memory and processing power are required, impacting system resources and potentially affecting overall performance.

Using locking protocols effectively is crucial for achieving high database performance while maintaining data consistency. Here are some tips to optimize the use of locking protocols for improved performance:

1. CHOOSE THE RIGHT ISOLATION LEVEL:

Select the isolation level that balances data consistency with performance. Lower isolation levels like Read Committed or Repeatable Read provide higher concurrency, while higher levels like Serializable offer stronger data consistency. Choose the appropriate level based on the application's requirements and acceptable trade-offs.

2. USE ROW-LEVEL LOCKING:

Whenever possible, opt for row-level locking instead of higher levels like page-level or table-level locking. Row-level locks minimize contention, allowing multiple transactions to access different parts of the same table simultaneously. This enhances concurrency and can lead to better performance.

3. AVOID OVERLY RESTRICTIVE LOCKS:

Be cautious of using overly restrictive locks, especially in read-heavy scenarios. Exclusive locks can cause contention and block other transactions, leading to decreased throughput. Use shared locks for read-only operations to enable multiple transactions to read simultaneously.

4. LIMIT LOCK DURATIONS:

Acquire locks only when necessary and release them as soon as they are no longer needed. Prolonged lock holding can increase contention and reduce concurrency. Minimize the time a transaction spends holding locks to maximize the number of transactions that can run concurrently.

5. BATCH SIMILAR OPERATIONS:

If possible, batch similar operations together to reduce the frequency of lock acquisitions and releases. For example, instead of updating rows individually, update them in a single batch. This reduces lock overhead and contention.

6. CONSIDER OPTIMISTIC CONCURRENCY CONTROL:

Explore optimistic concurrency control mechanisms in scenarios where conflicts are infrequent. Optimistic approaches can minimize locking overhead and contention, leading to improved performance. However, ensure the validation and retry mechanisms are well-tuned to prevent excessive retries.

7. MONITOR LOCKING BEHAVIOR:

Use monitoring tools to track and analyze locking behavior. Identify frequently locked resources, high contention areas, and potential deadlocks. This insight can help fine-tune locking strategies for better performance.

8. IMPLEMENT DEADLOCK DETECTION AND HANDLING:

Set up deadlock detection mechanisms to identify and resolve deadlocks promptly. While detection and resolution introduce overhead, the impact is often lower than letting deadlocks persist. Fine-tune deadlock timeout settings to minimize their impact on performance.

9. OPTIMIZE INDEXING:

Well-chosen indexes can reduce the need for locking by enabling more efficient data access paths. Indexes can help minimize the number of rows affected by a lock, reducing contention and improving performance.

10. REVIEW AND REFINE REGULARLY:

Continuously review and refine locking strategies based on changing application requirements, data access patterns, and performance benchmarks. Regularly analyzing performance metrics can help identify areas for improvement.

11. USE LOCK HINTS SPARINGLY:

Some DBMSs allow you to provide lock hints to query statements. While these hints can be useful in certain cases, use them sparingly and with caution. Incorrect use of lock hints can lead to unintended consequences and reduced performance.

12. BENCHMARK AND TEST:

Test your locking strategies under realistic workload scenarios. Conduct performance benchmarks to evaluate how different locking approaches impact concurrency, throughput, and response times. Use the results to fine-tune your locking strategy.

Incorporating these concepts into your database design and application development processes can help you effectively leverage locking protocols to achieve high performance while maintaining data consistency. The key is to strike a balance between data integrity and concurrency optimization, based on your application's specific requirements and usage patterns.

Locking protocols are essential for maintaining data integrity and consistency in a multi-user environment. However, their impact on performance can be significant due to factors such as contention, deadlock management, lock escalation, and concurrency limitations. Performance optimization involves selecting the appropriate locking protocol and granularity based on the application's requirements, access patterns, and trade-offs between data consistency and performance. Additionally, careful monitoring and tuning of lock-related parameters can help mitigate the potential performance impact of locking protocols.

LOCK ESCALATION:

Lock escalation is a mechanism where a DBMS converts many fine-grained locks into fewer coarse-grained locks to reduce overhead. While it can optimize performance by reducing the number of locks, it can also result in decreased concurrency. Proper tuning of the lock escalation threshold is essential to avoid adversely impacting performance and to ensure that contention remains manageable. Lock escalation is a mechanism employed in database management systems to manage and optimize the number of locks acquired during concurrent transactions. It involves converting multiple fine-grained locks into fewer coarse-grained locks to reduce overhead. Lock escalation can be useful in scenarios where a large number of fine-grained locks could lead to excessive overhead.

Here's an example of how lock escalation might be used:

SCENARIO: INVENTORY MANAGEMENT SYSTEM

Consider an inventory management system used by an online retail store. The system tracks the quantities of various products in the store's inventory. Each product is represented as a separate row in the "Products" table, and each row is protected by a row-level lock during updates.

THE PROBLEM:

As the store grows, the number of products in the inventory also increases, leading to a significant number of individual row-level locks being acquired and released. This could result in increased lock management overhead and potentially impact the overall system performance.

THE SOLUTION:

To address this issue, the database administrator decides to implement lock escalation. Instead of maintaining individual row-level locks for each product, the system will escalate the locks to page-level locks when a certain threshold is reached. This means that all the rows within a page (a group of contiguous rows) will be protected by a single lock.

IMPLEMENTATION STEPS:

1. **Setting a Threshold**:

   Determine the threshold at which lock escalation should occur. For instance, when the number of locks obtained on a page reaches a certain limit, the system will escalate the locks.

2. **Monitoring Lock Counts**:

   The DBMS continuously monitors the number of locks obtained on each page of the "Products" table. When the lock count on a page surpasses the threshold, the system initiates lock escalation.

3. **Lock Escalation Process**:

   Once lock escalation is triggered, the DBMS releases the individual row-level locks on that page and replaces them with a single page-level lock. This reduces the number of locks maintained by the system.

4. **Controlling Lock Granularity**:

   The system must ensure that escalating locks to a higher granularity does not compromise data integrity or lead to excessive contention. It's important to test and fine-tune the threshold to strike the right balance between overhead reduction and performance impact.

BENEFITS:

- **Reduced Lock Overhead**: Lock escalation reduces the overhead of managing numerous fine-grained locks, as the system now deals with a smaller number of coarse-grained locks.

- **Improved Concurrency**: With fewer locks to manage, the system can potentially support more concurrent transactions, enhancing concurrency and system responsiveness.
- **Performance Optimization**: By avoiding excessive lock management, the database can allocate more resources to actual data processing tasks, improving overall performance.
- **Resource Efficiency**: Lock escalation reduces the memory and processing resources required for lock management, making better use of system resources.

It's important to note that while lock escalation can be beneficial, it's not a one-size-fits-all solution. The threshold for lock escalation, the granularity of locks, and the specific scenarios where it's applied should be carefully considered and tested to ensure they align with the application's performance and concurrency requirements.

LOCK-BASED VS. LOCK-FREE CONCURRENCY CONTROL:

Lock-based concurrency control ensures data consistency but can lead to contention and reduced throughput. Lock-free approaches, such as optimistic concurrency control, allow transactions to proceed without acquiring locks, potentially leading to better performance. The choice between these approaches depends on the nature of the workload, the application's requirements, and the trade-offs between data consistency and performance.

MULTI-VERSION CONCURRENCY CONTROL (MVCC):

MVCC maintains multiple versions of data items to allow concurrent transactions to work on the same data without conflicts. This approach enhances performance by reducing contention and eliminating the need for explicit locking in many cases. However, maintaining multiple versions introduces storage overhead, and careful management of older versions is necessary to prevent excessive storage consumption.

Multi-Version Concurrency Control (MVCC) is a database technique that allows multiple transactions to access and modify the same data simultaneously while maintaining data consistency and isolation. It achieves this by creating multiple versions of data, each associated with a specific transaction or timestamp. Here's a brief example of how MVCC works:

SCENARIO: ONLINE BOOKING SYSTEM

Consider an online booking system where users can book flights. The system tracks available seats for each flight in a database. Multiple users can search for and book flights simultaneously, potentially leading to conflicts if not managed properly.

Using MVCC:

1. Initial State:

- Flight A has 50 available seats.
- Transaction T1 starts searching for available seats on Flight A.

2. Data Versioning:

- While T1 is searching, a new transaction T2 starts searching for available seats on the same flight.
- To avoid contention, MVCC creates a snapshot of the data at the start of each transaction. T1 and T2 each see their own version of the data.

3. Data Modification:

- T1 finds that there are 10 available seats and decides to book 5 of them.
- T2 finds that there are 10 available seats as well and decides to book 3 of them.

4. Data Commit:

- T1 commits its changes and releases its snapshot, making the changes visible to other transactions.
- T2 tries to commit its changes, but the system detects a conflict with T1's changes.

5. Conflict Resolution:

- MVCC resolves the conflict by ensuring that T2 sees the updated data after T1's changes.

- T2 modifies its changes to account for the seats booked by T1.

6. Final Result:

- Flight A now has 42 available seats (50 - 5 booked by T1 - 3 modified by T2).

BENEFITS OF MVCC:

- **Concurrency**: Both T1 and T2 were able to search and book seats concurrently, thanks to MVCC. They operated on their own versions of the data, avoiding contention and locking conflicts.

- **Data Consistency**: MVCC ensures that the final result is consistent by resolving conflicts and incorporating changes from different transactions. Users see a consistent and accurate view of the data.

- **Isolation**: Transactions operate in isolation with their own data versions, avoiding dirty reads, non-repeatable reads, and other anomalies.

- **Performance**: MVCC reduces the need for locking and contention, allowing for higher concurrency and better system performance.

In this example, MVCC allows multiple users to interact with the booking system concurrently without conflicts while maintaining data consistency. This technique is widely used in databases to provide high levels of concurrency and isolation for transactional systems.

CONCURRENCY CONTROL IN DISTRIBUTED DATABASES:

Distributed databases introduce additional challenges due to the geographical distribution of data. Concurrency control mechanisms must address network latencies, distributed transaction coordination, and global consistency. Achieving efficient concurrency control across distributed nodes requires a balance between data consistency and minimizing inter-node communication, which impacts performance.

IMPACT OF CONCURRENCY CONTROL ON PERFORMANCE:

The choice of concurrency control strategy directly impacts database performance. Aggressive locking may result in high contention, slowing down transactions, while looser controls may lead to anomalies but better throughput. Performance testing, profiling, and tuning are necessary to strike the right balance between concurrency and performance in specific use cases.

CONCURRENCY CONTROL AND INDEXING:

Indexing can impact concurrency and performance. Lock contention on frequently accessed rows can arise due to indexed columns, while missing indexes can lead to full table scans that hinder performance. Strategic indexing decisions based on query patterns and concurrency requirements can significantly improve both performance and data consistency. Indexes play a crucial role in minimizing the need for locking and reducing contention in a database. They provide efficient access paths to data, allowing transactions to retrieve specific rows without requiring extensive table scans. This efficient access helps mitigate the need for locks and reduces contention among transactions. Here's how indexes achieve this:

- **Quick Data Retrieval:** Indexes organize data in a way that enables the database to locate specific rows quickly. When a transaction needs to access certain data, it can use the index to directly navigate to the relevant rows, avoiding the need to scan the entire table. This efficient access reduces the time that a transaction needs to hold locks, minimizing contention.

- **Reduced Lock Scope:** With proper indexes in place, transactions can access and modify only the specific rows they need, rather than locking entire tables. This narrower lock scope reduces the chances of conflicts between transactions, as they are less likely to interfere with each other's operations.

- **Selective Locking:** Indexes allow for selective locking, where transactions can acquire locks on only the rows they intend to modify. This enables multiple transactions to work concurrently on different parts of the data without causing contention. Selective locking helps maintain data consistency while allowing for higher concurrency.

- **Covering Indexes:** Covering indexes include all columns needed to satisfy a query, effectively turning the index into a data storage structure itself. When a query can be fulfilled using a covering index, the database doesn't need to access the main table. This not only speeds up data retrieval but also reduces the need for locks on the base table, as the query operates solely within the index.
- **Index-Organized Tables (IOTs):** Some databases support index-organized tables, where the table data is physically stored within the index structure itself. This design eliminates the need for separate index and data storage, leading to more efficient data access and reduced locking contention.
- **Concurrency and Lock Modes:** Indexes can support different lock modes, allowing transactions to request shared locks for read operations and exclusive locks for write operations. The granularity provided by indexes allows for a fine-tuned approach to locking, minimizing the chances of conflicts and contention.

DEADLOCK AVOIDANCE:

Efficient access paths provided by indexes reduce the likelihood of transactions holding locks for extended periods. This decreases the chance of transactions waiting for each other and potentially leading to deadlocks. Consequently, index-based access paths contribute to smoother transaction execution and less contention.

Indexes provide efficient access paths to data, enabling transactions to retrieve the required information without excessive scanning and locking. By reducing the need for locks and providing selective, optimized data access, indexes help minimize contention among transactions. A well-designed indexing strategy can significantly enhance concurrency, improve performance, and contribute to maintaining data consistency in a multi-user database environment.

The intricate relationship between concurrency control and database performance underscores the need for a thoughtful and context-aware approach. Balancing data integrity, user experience, and query throughput requires understanding the intricacies of concurrency control mechanisms and their effects on various database operations. Striking this balance is a key aspect of the broader endeavor of optimizing database systems for performance.

CHAPTER 6. DATABASE PERFORMANCE TUNING TOOLS AND UTILITIES

Database performance tuning is a crucial aspect of ensuring optimal system performance and responsiveness. Performance tuning tools and utilities provide valuable assistance in diagnosing, analyzing, and optimizing database performance. This section covers several database performance tuning tools and utilities, highlighting their key features, benefits, and their role in empowering efficient database optimization.

MONITORING AND DIAGNOSTIC TOOLS:

1. **Database Profilers**: Profiling tools capture and analyze the behavior of database queries and transactions. They provide insights into query execution times, resource utilization, and bottlenecks. Profilers help identify slow queries, excessive resource consumption, and performance hotspots, facilitating targeted optimization efforts.

2. **Performance Monitoring Tools**: These tools monitor system-level performance metrics, including CPU utilization, memory usage, disk I/O, and network throughput. They help identify performance bottlenecks, track resource usage, and enable proactive performance management.

3. **Query Execution Analyzers**: Query execution analyzers examine query execution plans, identifying suboptimal access methods, join operations, or missing indexes. They provide visual representations of query plans, allowing for detailed analysis and identification of areas for query optimization.

4. **Wait Event Analyzers**: Wait event analyzers track and analyze database wait events, which occur when a process is waiting for a particular resource or event to complete. These tools help identify contention issues, performance bottlenecks, and resource-intensive operations, allowing for targeted performance tuning.

QUERY OPTIMIZATION TOOLS:

1. **SQL Profilers**: These profilers capture and analyze SQL statements executed by applications. They provide insights into query execution times, plan reuse, and overall query performance. SQL profilers help identify poorly performing queries, excessive round trips, or inefficient data access patterns.

2. **Query Rewriting Tools**: These tools assist in transforming queries into more efficient forms. They optimize query structures, eliminate unnecessary operations, and rewrite complex expressions to simplify query execution. These

tools automate the process of query optimization and enhance query performance.

3. **Index Analysis Tools**: These tools evaluate index usage, identify missing or redundant indexes, and suggest index optimizations. They analyze query patterns and execution plans to determine the most effective indexing strategies, aiding in improved query performance and reduced disk I/O.

OPTIMIZATION UTILITIES:

1. **Database Statistics Analyzers**: These utilities analyze database statistics and histograms to provide insights into data distribution, cardinality, and index selectivity. By maintaining accurate statistics, these tools enable the query optimizer to make better decisions and generate optimal execution plans.

2. **Automatic Performance Tuning**: Some database systems offer automatic performance tuning features that analyze query execution patterns, collect statistics, and automatically suggest performance improvements. These utilities automate the performance tuning process, reducing the need for manual intervention.

3. **Storage and Index Maintenance Utilities**: Storage and index maintenance utilities assist in optimizing storage structures, managing indexes, and performing data reorganization tasks. They help improve I/O performance, eliminate fragmentation, and ensure efficient data retrieval.

BENEFITS OF PERFORMANCE TUNING TOOLS AND UTILITIES:

➢ **Efficient Problem Identification**: Performance tuning tools and utilities enable comprehensive monitoring and analysis, allowing for efficient identification of performance bottlenecks, slow queries, and resource-intensive operations.

➢ **Targeted Optimization Efforts**: By providing detailed insights into query execution, resource utilization, and system performance, these tools enable targeted optimization efforts. They help prioritize optimization tasks based on the impact on performance.

➢ **Improved Decision-Making**: Performance tuning tools offer visual representations, analysis reports, and recommendations that empower database administrators and developers to make informed decisions regarding query optimization, index design, and overall system configuration.

➢ **Time and Resource Savings**: Automated performance tuning utilities streamline the optimization process, reducing the need for manual analysis and troubleshooting. They save time and effort, allowing for efficient utilization of resources.

Database performance tuning tools and utilities play a vital role in optimizing database performance, enhancing responsiveness, and improving overall system efficiency. By providing monitoring, diagnostic, query optimization, and optimization utility capabilities, these tools empower database administrators and developers to identify performance issues, optimize query execution, and fine-tune database configurations. Leveraging these tools enables proactive performance management, efficient problem identification, and targeted optimization efforts, ultimately leading to enhanced database performance, improved user experience, and optimized resource utilization.

Profiling and Tracing Techniques: Unveiling Insights for Performance Analysis

Profiling and tracing techniques are vital tools in performance analysis, helping developers and system administrators gain deep insights into the behavior and performance of software applications. These techniques involve capturing and analyzing data related to application execution, resource usage, and code behavior. This section covers profiling and tracing techniques, highlighting their significance in performance analysis and optimization.

PROFILING TECHNIQUES:

- **CPU Profiling**: CPU profiling focuses on analyzing the CPU usage of an application. It provides detailed information about the execution time of different functions and methods within the code. CPU profiling helps identify performance bottlenecks, hotspots, and areas of excessive computation, enabling developers to optimize code execution.

- **Memory Profiling**: Memory profiling aims to analyze an application's memory usage patterns. It helps identify memory leaks, excessive memory allocations, and inefficient memory management practices. By examining memory consumption at various levels, developers can optimize memory usage and improve overall application performance.

TRACING TECHNIQUES:

- Method Tracing: Method tracing involves capturing detailed information about method calls and their durations during application execution. It helps understand the sequence of method invocations, enabling the identification of performance issues related to method calls, such as excessive recursion or redundant invocations.

- Event Tracing: Event tracing captures events and interactions within an application, such as user actions, system events, or custom-defined events. It helps identify the sequence and timing of events, enabling analysis of performance-related dependencies, synchronization issues, or event-driven bottlenecks.

➢ Transaction Tracing: Transaction tracing focuses on tracking the execution path and performance of individual transactions or requests within an application. It provides insights into the flow of control, time spent in various components, and external dependencies. Transaction tracing facilitates the identification of performance bottlenecks specific to individual transactions.

BENEFITS AND USE CASES:

➢ **Performance Optimization**: Profiling and tracing techniques provide fundamental information for identifying performance bottlenecks and resource-intensive areas within an application. Developers can use this information to optimize pivotal code sections, reduce computational overhead, and improve overall application performance.

➢ **Resource Utilization Analysis**: Profiling and tracing enable analysis of resource utilization patterns, including CPU, memory, and I/O. By identifying resource-intensive operations, developers can optimize resource usage, reduce contention, and enhance system scalability.

➢ **Memory Leak Detection**: Memory profiling and tracing techniques help identify memory leaks, excessive allocations, or inefficient memory management practices. This enables developers to eliminate memory leaks, optimize memory usage, and enhance application stability.

➢ **Code Optimization**: Profiling and tracing provide valuable insights into the execution behavior of code. Developers can identify areas of redundant computations, inefficient algorithms, or slow-performing methods. This allows them to refactor and optimize code for improved performance and responsiveness.

➢ **Performance Tuning**: Profiling and tracing techniques aid in performance tuning by providing quantifiable metrics and data-driven analysis. Developers can measure the impact of optimizations, compare performance before and after changes, and fine-tune the application for optimal performance.

Profiling and tracing techniques are powerful tools in performance analysis, providing in-depth insights into the behavior, resource usage, and performance of software applications. By utilizing profiling techniques like CPU, memory, and I/O profiling, developers can identify performance bottlenecks, optimize code execution, and improve resource utilization. Tracing techniques, including method tracing, event tracing, and transaction tracing, enable analysis of code behavior, event sequences, and transaction performance. Leveraging these techniques, developers can optimize necessary sections, eliminate inefficiencies, and enhance the overall performance and responsiveness of applications. Profiling and tracing techniques are invaluable for performance analysis, optimization, and delivering high-quality software solutions.

MONITORING AND DIAGNOSTICS TOOLS:

Monitoring and diagnostics tools are essential components of effective performance management in software systems. These tools provide real-time and historical data about the behavior, health, and performance of applications and infrastructure. This section covers monitoring and diagnostics tools, highlighting their significance in proactively identifying issues, optimizing performance, and ensuring the smooth operation of software systems.

MONITORING TOOLS:

- **Application Performance Monitoring** (APM): APM tools monitor and measure the performance of software applications in real-time. They collect data on response times, throughput, error rates, and resource utilization. APM tools provide insights into application behavior, enabling early detection of performance issues, and facilitating proactive optimization.

- **Infrastructure Monitoring**: Infrastructure monitoring tools focus on monitoring the health and performance of servers, network devices, databases, and other infrastructure components. They track resource utilization, availability, and network latency, helping identify bottlenecks, capacity constraints, or hardware failures.

- **Log Monitoring and Analysis**: Log monitoring tools capture and analyze logs generated by applications, servers, or network devices. They help detect errors, exceptions, and anomalies, providing valuable insights into system behavior and troubleshooting issues. Log monitoring enables proactive identification and resolution of issues before they impact performance.

- **Synthetic Monitoring**: Synthetic monitoring tools simulate user interactions and transactions to assess application performance and availability. They generate artificial traffic and measure response times from multiple geographical locations. Synthetic monitoring enables businesses to proactively monitor end-user experience and identify performance degradation.

DIAGNOSTICS TOOLS:

1. **Profiling Tools**: Profiling tools capture detailed runtime information about code execution, resource utilization, and performance metrics. They help identify hotspots, bottlenecks, and resource-intensive sections of the code. Profiling tools assist in optimizing code, reducing computational overhead, and improving overall performance.

2. **Heap and Memory Analysis**: Heap and memory analysis tools provide insights into memory usage, object allocation, and memory leaks. They help identify excessive memory consumption, inefficient memory management, and memory-

related performance issues. These tools facilitate efficient memory optimization and improve application stability.

3. **Network Traffic Analysis**: Network traffic analysis tools capture and analyze network traffic, allowing for the identification of bandwidth constraints, latency issues, or network-related performance bottlenecks. They help pinpoint network anomalies, optimize network configurations, and enhance application performance.

4. **Database Performance Diagnostics**: Database performance diagnostics tools monitor and analyze database performance metrics, query execution plans, and resource utilization. They identify slow queries, index inefficiencies, and database configuration issues. Database performance diagnostics tools assist in optimizing database performance, improving query response times, and ensuring efficient data retrieval.

BENEFITS AND USE CASES:

- **Proactive Issue Detection**: Monitoring and diagnostics tools enable proactive identification of performance issues, anomalies, and errors. Early detection allows for timely resolution and minimizes the impact on system performance and user experience.

- **Performance Optimization**: By providing real-time and historical data about resource utilization, response times, and code behavior, these tools facilitate performance optimization efforts. Developers can identify bottlenecks, optimize requisite sections, and fine-tune system configurations for enhanced performance.

- **Scalability and Capacity Planning**: Monitoring tools help assess system scalability by tracking resource usage, response times, and workload patterns. They enable capacity planning by providing insights into resource demands and usage trends, helping organizations scale their infrastructure to meet growing demands.

- **Troubleshooting and Root Cause Analysis**: Diagnostics tools assist in troubleshooting performance issues and identifying root causes of system anomalies. They provide detailed data, logs, and metrics for in-depth analysis, accelerating the resolution of issues and minimizing system downtime.

- **SLA Compliance and Performance Benchmarking**: Monitoring tools allow organizations to monitor performance against service-level agreements (SLAs) and benchmark application performance against industry standards. They help ensure compliance with performance targets and enable businesses to assess their performance relative to competitors.

Monitoring and diagnostics tools are indispensable for efficient performance management, enabling organizations to monitor, analyze, and optimize the behavior and performance of software systems. By leveraging these tools, businesses can proactively identify issues, optimize necessary sections, troubleshoot anomalies, and ensure smooth application operation. Monitoring tools provide real-time insights into application and infrastructure performance, while diagnostics tools enable in-depth analysis of code execution, memory usage, and network behavior. Together, these tools empower organizations to deliver high-performing software, enhance user experience, and achieve optimal resource utilization. By embracing monitoring and diagnostics tools, businesses can stay ahead of performance challenges, maintain system reliability, and drive continuous improvement in their software systems.

PERFORMANCE TUNING BEST PRACTICES:

Optimizing Software Systems for Efficiency and Responsiveness:

Performance tuning is an essential aspect of ensuring efficient and responsive software systems. It involves identifying and resolving performance bottlenecks, optimizing resource utilization, and enhancing overall system performance. This section emphasizes performance tuning best practices that empower organizations to optimize their software systems and deliver exceptional user experiences.

- **Set Clear Performance Goals**: Establishing clear performance goals is the first step in performance tuning. Define specific performance targets, such as response time thresholds, throughput rates, or resource utilization limits. These goals provide a benchmark for evaluating and optimizing system performance.

- **Measure and Baseline**: Before initiating performance tuning efforts, establish a baseline by measuring the system's current performance. Collect relevant metrics such as response times, CPU and memory usage, database query times, and network latency. This baseline helps gauge the effectiveness of performance optimizations and serves as a reference point for future comparisons.

- **Identify Performance Bottlenecks**: Thoroughly analyze the system to identify performance bottlenecks. Utilize profiling tools, monitoring data, and diagnostic techniques to pinpoint areas of inefficiency. Common bottlenecks include slow database queries, inefficient algorithms, network congestion, and excessive CPU or memory usage.

- **Optimize Database Performance**: Database performance often has a significant impact on overall system performance. Implement effective indexing strategies, normalize database schemas, and optimize query execution plans. Regularly monitor and analyze database performance metrics to identify areas for improvement, such as index usage, query tuning, or data caching.

- **Efficient Resource Utilization**: Optimize resource utilization to maximize system efficiency. Avoid resource contention by appropriately configuring hardware, network, and software components. Fine-tune memory allocation, thread management, and I/O operations to ensure optimal utilization of CPU, memory, and disk resources.

- **Code Optimization**: Optimize code to improve performance and reduce computational overhead. Eliminate unnecessary loops, redundant calculations, and excessive object instantiation. Utilize efficient data structures and algorithms. Profile the code to identify hotspots and focus optimization efforts on pivotal sections.

- **Caching and Data Retrieval Optimization**: Implement caching mechanisms to store frequently accessed data and avoid redundant computations. Leverage in-memory caching, database query caching, or content delivery networks (CDNs) to reduce the load on backend systems and improve response times. Optimize data retrieval operations by fetching only required data and minimizing data transfer.

- **Parallelism and Concurrency**: Leverage parallel processing and concurrency techniques to optimize performance. Utilize multi-threading, asynchronous operations, or distributed computing to parallelize tasks and maximize resource utilization. Be mindful of synchronization and resource-sharing issues to avoid contention and bottlenecks.

- **Load Testing and Performance Benchmarking**: Conduct thorough load testing and performance benchmarking to assess system performance under various scenarios and stress levels. Simulate realistic workloads, user loads, and data volumes to identify system limitations and ensure that performance goals are met. Continuously monitor and measure performance to detect regressions and validate optimizations.

- **Continuous Monitoring and Optimization**: Performance tuning is an ongoing process. Implement comprehensive monitoring and diagnostics to continuously assess system performance and detect anomalies. Regularly review performance metrics, logs, and user feedback to identify areas for further optimization. Embrace a culture of continuous improvement to ensure that system performance evolves with changing requirements.

Performance tuning best practices provide a roadmap for optimizing software systems and delivering exceptional user experiences. By setting clear performance goals, measuring baselines, and identifying bottlenecks, organizations can focus their efforts on the most crucial areas. Optimizing database performance, efficient resource utilization, code optimization, caching strategies, and parallelism techniques enhance system responsiveness and scalability. Regular load testing and benchmarking validate performance improvements and ensure that system performance meets expectations. Continuous monitoring and optimization guarantee that software systems evolve and adapt to changing demands. By embracing performance tuning best practices, organizations can achieve efficient, high-performing software systems that exceed user expectations and drive business success.

CHAPTER 7. HARDWARE AND INFRASTRUCTURE CONSIDERATIONS

Hardware and infrastructure considerations play a crucial role in optimizing database performance. The hardware components and infrastructure on which a database system operates can significantly impact its speed, scalability, and overall efficiency. Factors such as server capacity, disk I/O, memory, network bandwidth, and storage configurations directly influence the database's ability to handle and process data efficiently. This chapter highlights the importance of hardware and infrastructure considerations in database optimization and sets the stage for exploring key considerations and best practices in this area.

Storage Systems and Disk I/O Optimization: Key Considerations for Database Performance Tuning

Storage systems and disk I/O optimization are vital aspects of performance tuning and optimization for databases. The storage infrastructure and disk input/output (I/O) operations significantly impact the overall performance, responsiveness, and scalability of database systems. This section delves into the importance of storage systems and disk I/O optimization in database performance tuning, highlighting key considerations and best practices to enhance efficiency and maximize the database's capabilities.

STORAGE SYSTEMS AND DISK I/O OPTIMIZATION:

- **Speed and Responsiveness**: Efficient storage systems and optimized disk I/O operations directly impact the speed and responsiveness of database transactions. Faster disk I/O operations reduce the time required to read and write data, resulting in improved query response times and enhanced user experiences.

- **Scalability and Concurrency**: Storage systems and disk I/O optimization are crucial for scaling database systems to handle increasing data volumes and concurrent user loads. By minimizing disk I/O bottlenecks and optimizing data retrieval mechanisms, organizations can support a larger number of users and transactions without sacrificing performance.

- **Data Integrity and Reliability**: Reliable storage systems and optimized disk I/O operations are essential for ensuring data integrity and minimizing the risk of data loss or corruption. Proper storage configurations, including redundancy, backup mechanisms, and fault-tolerant setups, enhance data reliability and protect against potential failures.

KEY CONSIDERATIONS FOR STORAGE SYSTEMS AND DISK I/O OPTIMIZATION:

- **Storage Medium Selection:** Choose the appropriate storage medium based on the specific needs of the database. Solid-state drives (SSDs) offer faster access times and lower latency compared to traditional hard disk drives (HDDs). Analyze the workload requirements, cost considerations, and expected performance gains to determine the optimal storage medium.

- **RAID Configuration**: Implement Redundant Array of Independent Disks (RAID) configurations to enhance data availability, fault tolerance, and I/O performance. RAID levels, such as RAID 0, RAID 1, RAID 5, or RAID 10, offer different combinations of data striping, mirroring, and parity for improved performance and data redundancy.

- **Disk Partitioning and Alignment**: Proper disk partitioning and alignment ensure optimal disk I/O performance. Align partitions with the underlying storage system's physical block size to avoid unnecessary read-modify-write operations. Allocate separate partitions for data files, log files, and operating system files to minimize contention and optimize disk access.

- **I/O Scheduler Configuration**: Tune the I/O scheduler settings to optimize disk I/O operations based on the workload characteristics. Different I/O scheduler algorithms, such as CFQ (Completely Fair Queuing), NOOP, or Deadline, have varying performance impacts. Analyze the workload pattern and choose the appropriate scheduler for optimal disk I/O performance.

- **Read and Write Caching**: Utilize read and write caching mechanisms to improve disk I/O performance. Implementing caching at various levels, such as database buffer cache, file system cache, or operating system-level caching, reduces disk access frequency and improves response times for frequently accessed data.

- **File System Selection**: Selecting an appropriate file system can impact disk I/O performance. File systems like ext4, XFS, or NTFS have different characteristics and capabilities. Consider factors such as file system journaling, metadata handling, and scalability to choose the optimal file system for the database workload.

- **Monitoring and Performance Tuning**: Regularly monitor disk I/O metrics, such as throughput, latency, and queue lengths, to identify performance bottlenecks and optimize disk I/O operations. Use monitoring tools and performance tuning techniques to fine-tune parameters like block size, I/O concurrency, and buffer cache size for optimal disk I/O performance.

Storage systems and disk I/O optimization are fundamental considerations in performance tuning and optimization for databases. By optimizing disk I/O operations, organizations can enhance database speed, responsiveness, scalability, and data reliability. Considering factors such as storage medium selection, RAID configurations, disk partitioning, I/O scheduler settings, caching mechanisms, file system choices, and continuous monitoring, organizations can unlock the full potential of their database systems. Implementing best practices in storage systems and disk I/O optimization ensures efficient data access, improves user experiences, and maximizes the performance and efficiency of database applications.

Memory Management and Caching Strategies: Key Considerations for Database Optimization

Memory management and caching strategies play a pivotal role in optimizing and fine-tuning databases for improved performance and responsiveness. Efficient memory utilization and effective caching mechanisms can significantly enhance query execution times, reduce disk I/O operations, and enhance overall system performance. This section emphasizes the importance of memory management and caching strategies in database optimization and highlights key considerations to achieve optimal performance.

MEMORY MANAGEMENT AND CACHING STRATEGIES:

- **Query Response Time**: Effective memory management and caching strategies minimize disk I/O operations by storing frequently accessed data in memory. This reduces the time required to retrieve data from disk, leading to faster query response times and improved overall system performance.

- **Scalability and Concurrency**: Optimal memory management and caching allow database systems to efficiently handle increasing workloads and concurrent user requests. By utilizing memory effectively, organizations can support a larger number of users and transactions without sacrificing performance.

- **Resource Utilization**: Efficient memory management reduces the need for frequent disk access, resulting in reduced resource contention and improved resource utilization. This allows the database to handle more concurrent requests and optimize the utilization of other system resources, such as CPU and disk I/O.

KEY CONSIDERATIONS FOR MEMORY MANAGEMENT AND CACHING STRATEGIES:

- **Buffer Pool Management**: The buffer pool, a dedicated portion of memory, holds frequently accessed data pages in memory. Proper configuration of the buffer pool size is essential to ensure optimal memory usage. Allocate sufficient memory to the buffer pool to accommodate frequently accessed data, minimizing disk I/O and enhancing performance.

- **Query Cache**: Implement a query cache mechanism to store and reuse the results of frequently executed queries. This eliminates the need to re-execute identical queries, reducing the processing overhead and improving response times. However, careful consideration should be given to the query cache size and cache invalidation policies to ensure data consistency.

- **Index and Metadata Caching**: Caching frequently accessed indexes and metadata in memory can significantly improve query performance. By keeping frequently used index pages and metadata in memory, database systems can avoid disk access for index lookups and minimize the impact of disk I/O on query execution.

- **Data Caching**: Implement data caching mechanisms, such as in-memory databases or distributed caching frameworks, to store frequently accessed data in memory. This allows for faster data retrieval and reduces the need for disk I/O operations. Considerations should be given to cache eviction policies, cache consistency, and cache synchronization mechanisms.

- **Eviction Policies**: Define appropriate eviction policies for caching mechanisms to ensure that the most relevant and frequently accessed data remains in memory. LRU (Least Recently Used) and LFU (Least Frequently Used) are common eviction policies that can be employed based on the specific application requirements and data access patterns.

- **Memory Allocation and Configuration**: Optimize memory allocation and configuration parameters based on the database system and workload characteristics. Consider factors such as memory allocation for sorting and joining operations, buffer sizes, memory caching limits, and memory-related system parameters. Regularly monitor memory usage and adjust configuration settings accordingly.

- **Monitoring and Performance Tuning**: Continuously monitor memory usage, cache hit rates, and overall system performance to identify potential bottlenecks or areas for optimization. Use performance tuning techniques, such as profiling and query analysis, to fine-tune memory management and caching strategies. Regularly review and adjust memory-related configuration parameters to adapt to changing workload requirements.

Memory management and caching strategies are instrumental in optimizing database performance, reducing disk I/O operations, and enhancing system responsiveness. By effectively managing memory resources, leveraging caching mechanisms, and implementing proper eviction policies, organizations can minimize data retrieval time, improve query response times, and handle larger workloads with greater concurrency. Key considerations, such as buffer pool management, query caching, index and metadata caching, data caching, eviction policies, memory allocation, and continuous monitoring, ensure optimal memory utilization and efficient caching. By implementing these strategies, organizations can achieve high-performance database systems, providing faster query execution, improved user experiences, and enhanced overall system efficiency.

CPU UTILIZATION AND PARALLEL PROCESSING:

CPU utilization and parallel processing are essential factors in optimizing and fine-tuning databases for improved performance and responsiveness. Efficient utilization of CPU resources and leveraging parallel processing techniques can significantly enhance query execution, reduce response times, and improve overall system performance. This section emphasizes the importance of CPU utilization and parallel processing in database optimization and highlights key considerations to achieve optimal performance.

IMPORTANCE OF CPU UTILIZATION AND PARALLEL PROCESSING:

- **Query Execution Speed**: Optimizing CPU utilization is crucial for speeding up query execution. By efficiently utilizing CPU resources, organizations can process queries faster, resulting in reduced response times and improved overall system performance.

- **Concurrency and Scalability**: Leveraging parallel processing techniques allows database systems to handle concurrent user requests and larger workloads effectively. Distributing query processing across multiple CPU cores enables simultaneous execution of multiple queries, enhancing system scalability and performance.

- **Resource Optimization**: Efficient CPU utilization helps optimize system resources, including power consumption and heat generation. By utilizing CPU resources effectively, organizations can maximize the throughput of their database systems and minimize wasted resources.

KEY CONSIDERATIONS FOR CPU UTILIZATION AND PARALLEL PROCESSING:

- **Query Optimization**: Efficient query optimization plays a crucial role in CPU utilization. Analyze query execution plans, identify resource-intensive operations, and optimize queries to minimize CPU usage. Techniques such as proper indexing, query rewriting, and analyzing execution plans help reduce CPU-intensive operations.

- **Parallel Query Execution**: Leverage parallel query execution to distribute query processing across multiple CPU cores. Configure the database system to execute queries in parallel, allowing simultaneous processing of multiple queries and optimal utilization of CPU resources.

- **Parallel Indexing**: Parallelize index creation and maintenance tasks to speed up index operations and reduce the impact on CPU resources. Distributing the workload across multiple CPU cores allows for faster indexing, improving query performance.

➢ **Resource Allocation and Configuration**: Allocate adequate CPU resources to the database system to ensure optimal performance. Consider factors such as CPU cores, CPU affinity settings, and CPU scheduling to allocate resources effectively and avoid resource contention.

➢ **Load Balancing**: Implement load balancing mechanisms to distribute query processing across multiple servers or CPU cores. Load balancing ensures equal distribution of workload, preventing overutilization of specific CPUs and maximizing overall system performance.[1]

➢ **Multithreading and Asynchronous Operations**: Utilize multithreading and asynchronous programming techniques to leverage parallelism within the application code. By allowing concurrent execution of multiple tasks, organizations can improve CPU utilization and reduce idle time, leading to enhanced performance.

➢ **Monitoring and Performance Tuning**: Regularly monitor CPU utilization, query execution times, and system performance to identify potential bottlenecks or areas for optimization. Use performance tuning techniques, such as profiling and query analysis, to fine-tune CPU utilization and parallel processing strategies. Adjust database and application configurations based on monitoring results to optimize CPU usage.

Efficient CPU utilization and effective parallel processing techniques are key for optimizing database performance and achieving superior system responsiveness. By optimizing CPU utilization, organizations can reduce query response times, handle larger workloads, and maximize system scalability. Key considerations, including query optimization, parallel query execution, parallel indexing, resource allocation, load balancing, multithreading, and continuous monitoring, ensure optimal CPU utilization and effective parallel processing. By implementing these strategies, organizations can achieve high-performance database systems, deliver faster query execution, improve user experiences, and enhance overall system efficiency.

NETWORK AND LATENCY CONSIDERATIONS:

Network and latency considerations are essential elements in optimizing and fine-tuning databases for optimal performance and responsiveness. The efficiency of network communication and reduction of latency play a critical role in ensuring fast data retrieval, timely query processing, and seamless user experiences. This section highlights the importance of network and latency considerations in database optimization and explores key factors and considerations to achieve optimal performance.

IMPORTANCE OF NETWORK AND LATENCY CONSIDERATIONS:

➢ **Data Transfer Speed**: Efficient network communication enables fast data transfer between the application layer and the database layer. Optimized network performance ensures that data can be transmitted swiftly, reducing the time required for query execution and enhancing overall system performance.

➢ **Query Response Times**: Reducing network latency directly contributes to faster query response times. Minimizing the time taken to send and receive data between the client and the database server allows for quicker processing and retrieval of query results, improving user experiences.

➢ **Scalability and Availability**: Efficient network infrastructure supports system scalability and availability. A well-designed network architecture enables the distribution of database resources across multiple servers, facilitating load balancing and providing high availability for users, even under heavy workloads.

KEY CONSIDERATIONS FOR NETWORK AND LATENCY OPTIMIZATION:

➢ **Network Architecture**: Design a robust network architecture that accommodates the specific needs of the database system. Consider factors such as network topology, bandwidth capacity, and network hardware configurations. Redundancy and fault-tolerant mechanisms, such as redundant network connections and failover configurations, ensure reliable network communication.

➢ **Bandwidth and Throughput**: Ensure sufficient bandwidth and throughput to handle the expected workload and data transfer requirements. Monitor network utilization and upgrade network infrastructure if necessary to prevent bottlenecks and maintain optimal network performance.

➢ **Latency Reduction**: Identify and reduce network latency to minimize the delay in data transmission. Techniques such as optimizing network protocols, reducing network hops, and using compression techniques for data transfer can help reduce latency and enhance overall performance.

➢ **Network Security**: Implement robust network security measures to protect data during transmission. Secure network protocols, encryption mechanisms, and firewalls ensure data integrity and prevent unauthorized access, while maintaining optimal network performance.

➢ **Network Monitoring and Analysis**: Regularly monitor and analyze network performance metrics, such as latency, packet loss, and network congestion. Utilize network monitoring tools to identify bottlenecks, detect performance degradation, and troubleshoot network issues promptly. Analyzing network data can provide insights into areas that require optimization and ensure optimal network performance.

➢ **Data Compression and Optimization**: Implement data compression techniques to reduce data size during network transmission. Compressed data requires less bandwidth, resulting in faster data transfer and reduced network latency. Utilize compression algorithms and protocols suited to the specific database system and network environment.

➢ **Network Protocol Optimization**: Optimize network protocols and configurations to minimize overhead and reduce latency. Consider protocols such as TCP/IP tuning, adjusting buffer sizes, and optimizing network stack configurations to maximize network throughput and minimize latency.

➢ **Data Center Proximity**: If the database system is distributed across multiple data centers, consider the physical proximity between them. Reducing the distance between data centers can reduce network latency, enabling faster data replication and communication between distributed database components.

Network and latency considerations are fundamental for optimizing database performance, reducing query response times, and delivering seamless user experiences. Efficient network infrastructure, reduced latency, and optimized network protocols contribute to faster data transfer, timely query processing, and improved system scalability. By considering factors such as network architecture, bandwidth, latency reduction techniques, network security, monitoring, and optimization, organizations can achieve optimal network performance and enhance the overall efficiency and responsiveness of their database systems. Embracing these considerations enables organizations to provide efficient data retrieval, minimize query response times, and ensure a smooth user experience, even under demanding workloads.

CHAPTER REFERENCES

(1) K, S.; S, S.; P, Y.; K, G. Library Management System. *IJRASET* **2023**, *11* (8), 992–1011. https://doi.org/10.22214/ijraset.2023.55279.

CHAPTER 8. PERFORMANCE TUNING FOR SPECIFIC DATABASE SYSTEMS

Performance tuning is an essential aspect of optimizing specific database systems for enhanced efficiency and responsiveness. Regardless of the specific database brand, the principles and considerations for performance tuning remain largely consistent. Key areas of focus include query optimization, index tuning, memory management, disk I/O optimization, and network configuration.

Query optimization plays a central role in improving performance. Analyzing and fine-tuning query execution plans, ensuring proper indexing, and minimizing unnecessary data retrieval are essential steps. By optimizing queries, organizations can significantly reduce response times and enhance overall system performance.

Index tuning is another vital aspect of performance tuning. By carefully selecting and creating indexes on appropriate columns, organizations can improve data retrieval efficiency and accelerate query execution. Regular maintenance of indexes, including reorganization and rebuilding, helps maintain optimal performance as the database grows.

Effective memory management is key for performance tuning. Allocating sufficient memory to critical operations, optimizing buffer cache sizes, and ensuring efficient memory utilization can reduce disk I/O operations, improve query response times, and enhance overall system efficiency.

Disk I/O optimization involves minimizing disk access and maximizing data retrieval speed. Techniques such as appropriate disk partitioning, proper file system configurations, caching mechanisms, and intelligent I/O scheduling help reduce latency and optimize data retrieval, improving database performance.

Optimizing network configuration and communication is vital for efficient data transfer between clients and the database server. Reducing network latency, optimizing network protocols, and ensuring adequate bandwidth are key considerations. By minimizing network overhead and maximizing data transfer rates, organizations can enhance overall system responsiveness and user experience.

Performance tuning for specific database systems encompasses query optimization, index tuning, memory management, disk I/O optimization, and network configuration. Applying these principles, regardless of the specific database brand, helps organizations optimize performance, reduce response times, and ensure efficient and reliable database operations.

Database	Query Optimization	Database Performance
Oracle	SQL Developer SQL Plus SQL Tuning Advisor Oracle Performance Analyzer Diagnostic and Tuning Pack Oracle Trace EXPLAIN statement	Oracle Enterprise Manager (OEM) Oracle Grid Control Automatic Workload Repository (AWR) Oracle Statspack Trace File Analyzer Toad SolarWinds DPA
MySQL	MySQL Performance Schema MySQL Query Analyzer EXPLAIN statement MySQL Tuner Pt-query-digest EXPLAIN statement	MySQL WorkBench Percona Monitoring and Management MySQL Enterprise Audit MySQL Replication
MS SQL Server	SQL Server Query Analyzer SQL Server Query Store Database Tuning Advisor (DTA) Plan Explorer EXPLAIN statement	SQL Server Management Studio (SSMS) SQL Server Data Collector SQL Server Extended Events SQL Server Profiler
PostgreSQL	EXPLAIN statement ANALYZE statement pg_hint_plan PgTAP	pgAdmin pg_stat_statements pg_stat_activity pg_stat_monitor PgBadger pg_log PgBouncer Pgpool-II
MongoDB	db.collection.explain() db.setProfilingLevel()	MongoDB Compass mongotop Mongostat MongoDB Profiler MongoDB Logs

THE EXPLAIN STATEMENT

The 'EXPLAIN' statement, which is used for query plan analysis. The 'EXPLAIN' statement generally allows you to examine the execution plan that the DBMS query planner generates for a given SQL statement. This plan provides insights into how the DBMS intends to retrieve the data, which indexes it plans to use, and the order in which it will perform various operations like joins and sorting.

To use the 'EXPLAIN' statement in SQLite, you can simply prefix your SQL statement with `EXPLAIN`. For example:

```
EXPLAIN SELECT * FROM your_table WHERE your_condition;
```

When you execute this 'EXPLAIN' statement, the DBMS will return a result set containing information about the query execution plan. You can use this information to assess the efficiency of your query and identify potential areas for optimization.

Keep in mind that each DBMS's 'EXPLAIN' output may look different from that of other database management systems, as each DBMS uses its own terminology and format for query plan explanations. However, it provides valuable insights into how the DBMS plans to execute your query, which can be useful for query optimization and performance tuning.

Here is a summary of how the EXPLAIN statement works with different database management systems:

- **Oracle**: uses the keyword `EXPLAIN PLAN` followed by a SQL statement to display the execution plan of that statement. The output is stored in a table named `PLAN_TABLE` by default, which can be queried later using the `DBMS_XPLAN` package. The output shows the hierarchy of operations in the execution plan, along with their costs, cardinalities, and other details.

- **MySQL**: uses the keyword `EXPLAIN` followed by a SQL statement to display the execution plan of that statement2. The output is a table that shows one row for each table access operation in the execution plan, along with their types, costs, keys, and other information. The output can also be formatted as JSON or XML using the `FORMAT` option.

- **MS SQL Server**: uses the keyword `EXPLAIN` to show the query plan for an Azure Synapse Analytics SQL statement without running the statement. The output is an XML document that lists all operations in the query plan, each enclosed by the `<dsql_operation>` tag. The output also includes cost information for the query and recommendations to optimize the SQL statement performance.

- **PostgreSQL**: uses the keyword `EXPLAIN` followed by an optional list of options and a SQL statement to display the query plan of that statement. The output is a tree-like structure that shows the hierarchy of operations in the query plan, along with their estimated costs, row counts, and other details. The output format can be plain text, JSON, XML, or YAML.

- **IBM DB2**: uses the keyword `EXPLAIN` followed by a SQL statement to obtain information about access path selection for an explainable statement . A statement is explainable if it is a SELECT, MERGE, TRUNCATE, or INSERT statement, or the searched form of an UPDATE or DELETE statement . The information that is obtained is placed in a set of supplied user tables that are called EXPLAIN tables .[1]

- **Elasticsearch**: does not have a native EXPLAIN statement at the time of this writing. You can make queries to Elasticsearch using its REST API through the `_search` endpoint. When you make a query, you can request an explanation of how the scoring works for each document matching the query by setting the `explain` parameter to `true` in your query. This can help you understand why certain documents are ranked higher than others in your search results.

- **SQLite**: supports the `EXPLAIN` statement, which is used for query plan analysis. The `EXPLAIN` statement in SQLite allows you to examine the execution plan that the SQLite query planner generates for a given SQL statement. The `EXPLAIN QUERY PLAN` keyword shows a high-level overview of how SQLite will run the SQL statement.

- **MariaDB**: MariaDB uses the keyword `EXPLAIN` followed by a SQL statement to display the execution plan of that statement. The output is similar to MySQL's output, but it also includes some additional information such as extra columns, optimizer trace, and histograms.

- **Pinecone**: Pinecone is a cloud-based service for building and deploying vector similarity search applications. Pinecone does not use SQL and does not have a native EXPLAIN statement, but it has a method called `explain_query()` that returns an explanation of how Pinecone ranks documents for a given query vector. The output is a list of tuples that contain document IDs, scores, and explanations. The explanations are based on cosine similarity between the query vector and document vectors in Pinecone's vector index.

- **MongoDB**: MongoDB is a NoSQL database that uses a document-oriented data model and a JSON-like query language. It uses the method `explain()` on a cursor object to return information on how MongoDB executes a query. The output is a document that contains various fields describing the query plan, such as `queryPlanner`, `executionStats`, and `serverInfo`. The output also includes information on indexes, stages, and execution time.

- **Redis**: Redis is an in-memory data store that primarily uses simple data structures like strings, lists, and sets. Redis does not support SQL queries and does not have a native EXPLAIN statement, but it has a module called `redis-explain` that provides a similar functionality. The module uses a command `EXPLAIN` followed by a Redis command to show how the command is executed internally by Redis. The output is a list of steps that describe the data structures, algorithms, and operations involved in executing the command.

ORACLE PERFORMANCE TOOLS

Oracle Database offers a range of performance optimization, monitoring, and diagnostic tools to help database administrators (DBAs) manage and tune the database for optimal performance. These tools can be classified into several general groups based on their functionality:

1. **Database Management Tools:**

   ➢**Oracle Enterprise Manager (OEM):** OEM provides a comprehensive, web-based interface for managing and monitoring Oracle databases. It offers performance tuning advisors, automated diagnostics, and a central dashboard for monitoring the entire database environment.

   ➢**SQL Developer:** This integrated development environment (IDE) includes performance tuning features such as SQL tuning advisors, SQL profiling, and the ability to generate and analyze execution plans.

   ➢**SQLPlus:** Oracle's command-line interface provides basic monitoring and reporting capabilities, allowing DBAs to run SQL scripts for performance analysis.

2. **Performance Monitoring Tools:**

   ➢**Oracle Grid Control:** Part of OEM, Grid Control provides advanced monitoring capabilities for large database environments, offering real-time performance metrics, alerts, and historical data analysis.

   ➢**Oracle Automatic Workload Repository (AWR):** AWR collects and retains performance statistics for database analysis. It's the foundation for many Oracle performance monitoring and diagnostic tools.

   ➢**Oracle Statspack:** Statspack is a simpler alternative to AWR, providing detailed performance snapshots and reports for performance analysis.

3. **Performance Tuning Tools:**

   ➢**Oracle SQL Tuning Advisor:** SQL Tuning Advisor offers automated SQL statement tuning recommendations, including index recommendations, restructuring SQL, and gathering statistics.

   ➢**Oracle Automatic SQL Tuning:** This feature automates the process of identifying and tuning poorly performing SQL statements using SQL profiles and plan evolution.

➢**Oracle Performance Analyzer:** Part of OEM, this tool analyzes and optimizes database and SQL performance through the use of database replay, real application testing, and SQL tuning sets.

4. **Diagnostic Tools:**

➢**Oracle Diagnostics Pack:** This pack includes tools like Oracle Diagnostician and Oracle Trace File Analyzer (TFA) for advanced diagnostic and troubleshooting capabilities.

➢**Oracle Database Diagnostic and Tuning Packs:** These packs include features like SQL Access Advisor, SQL Tuning Sets, and Database Replay for in-depth performance diagnostics.

5. **Utilities:**

➢**Oracle Trace:** Oracle Trace provides detailed tracing and performance monitoring capabilities, enabling DBAs to capture and analyze SQL execution and wait events.

➢**Oracle Trace File Analyzer (TFA):** TFA helps in diagnosing and resolving issues by collecting and analyzing diagnostic data from Oracle databases and other components.

➢**Explain Plan:** Oracle's `EXPLAIN PLAN` statement and graphical tools allow DBAs to analyze execution plans for SQL statements, helping in query optimization.

6. **Third-Party Tools:** Additionally, there are many third-party tools and monitoring solutions available for Oracle Database that offer a wide range of performance optimization and diagnostic features. There are several third-party tools and monitoring solutions available for Oracle Database, each offering a unique set of features for performance optimization, monitoring, and diagnostics. Here are some popular ones:

- **SolarWinds Database Performance Analyzer:** SolarWinds provides a comprehensive database performance monitoring tool that supports Oracle Database, offering real-time monitoring, historical analysis, query tuning, and alerting features.

- **Quest Toad for Oracle:** Toad for Oracle is a widely used database management and development tool that includes performance tuning features such as SQL optimization, SQL monitoring, and database diagnostics.

- **Redgate SQL Monitor:** Redgate SQL Monitor offers real-time performance monitoring and alerting for Oracle Database, including historical query performance analysis and customizable dashboards.

- **Idera Database Performance Monitor (formerly VividCortex):** Idera's Database Performance Monitor provides deep insights into database performance, including query analysis, resource utilization, and anomaly detection, for Oracle and other database systems.

- **DBmaestro Database DevOps Platform:** DBmaestro offers database release automation and DevOps tools for Oracle Database, helping organizations streamline database changes while ensuring performance and compliance.

- **New Relic Database Monitoring:** New Relic provides a cloud-based monitoring solution that supports Oracle Database, offering application and infrastructure performance monitoring, as well as query analysis capabilities.

- **AppDynamics Database Monitoring:** AppDynamics offers application performance monitoring with built-in database monitoring features, including Oracle Database, for end-to-end visibility into application performance and database health.

- **SQL Diagnostic Manager for SQL Server and Oracle:** SQL Diagnostic Manager by IDERA provides real-time monitoring and alerting for Oracle Database, offering query analysis and performance tuning recommendations.

- **Dell Spotlight on Oracle:** Dell's Spotlight on Oracle is a database monitoring and performance optimization tool that provides real-time diagnostics and query analysis for Oracle Database.

- **Oracle SQL Developer:** While not a third-party tool, Oracle SQL Developer is a free integrated development environment (IDE) for Oracle Database that includes performance tuning features like SQL optimization and query execution plans.

These third-party tools and monitoring solutions enhance the capabilities of Oracle Database administrators and developers, enabling them to effectively manage, monitor, and optimize the performance of Oracle Database systems.

MYSQL PERFORMANCE TOOLS

MySQL offers various performance optimization, monitoring, and diagnostic tools to help database administrators and developers manage and tune MySQL database systems. These tools can be classified into several general groups based on their functionality:

1. **Database Management Tools:**

    - **MySQL Workbench:** MySQL Workbench is a comprehensive GUI tool that provides database design, administration, and performance monitoring capabilities. It includes tools for query optimization, performance tuning, and database design.

2. **Performance Monitoring and Tuning Tools:**

   ➢ **MySQL Enterprise Monitor:** This is a commercial tool by Oracle that provides advanced monitoring, alerts, and advisory services for MySQL databases. It offers real-time performance monitoring, query analysis, and recommendations for query optimization.

   ➢ **Percona Monitoring and Management (PMM):** PMM is an open-source platform for managing and monitoring MySQL databases. It includes features like query analytics, real-time monitoring, and customizable dashboards.

   ➢ **MySQL Performance Schema:** Performance Schema is a built-in MySQL feature that provides detailed performance instrumentation for monitoring and diagnosing database performance. It's a powerful tool for query analysis and optimization.

   ➢ **MySQL Query Analyzer:** Query Analyzer is part of MySQL Enterprise Edition and helps identify and analyze slow queries for optimization.

3. **Query Analysis and Optimization Tools:**

   ➢ **EXPLAIN Statement:** MySQL provides the 'EXPLAIN' statement, which helps analyze and optimize query execution plans. It shows how MySQL executes a query, which can be invaluable for improving query performance.

   ➢ **MySQL Tuner:** MySQL Tuner is a Perl script that analyzes MySQL performance and provides recommendations for configuration changes to optimize performance.

   ➢ **pt-query-digest:** This tool is part of the Percona Toolkit and helps analyze MySQL query performance by processing MySQL slow query logs.

4. **Diagnostic Tools:**

   ➢ **MySQL Enterprise Backup:** This commercial tool by Oracle provides backup and recovery features, allowing you to restore databases in case of issues.

   ➢ **MySQL Enterprise Audit:** Also a commercial tool by Oracle, this enables auditing and compliance features, helping in diagnosing and tracking database activity.

5. **Replication and Clustering Tools:**

- **MySQL Replication:** MySQL provides built-in replication features for creating replicas of the database. Tools like MySQL Shell and MySQL Router help manage replication and clustering configurations.
- **Galera Cluster:** Galera Cluster is an open-source synchronous multi-master replication solution for MySQL that offers high availability and scalability. It includes management tools like Galera Manager.

6. **Third-Party Tools:** There are numerous third-party monitoring and diagnostic tools available for MySQL, including Zabbix, Nagios, and Datadog, which offer additional features and integrations for MySQL monitoring.

MICROSOFT SQL SERVER PERFORMANCE TOOLS

Microsoft SQL Server offers a range of performance optimization, monitoring, and diagnostic tools to help database administrators manage and tune SQL Server database systems. These tools can be classified into several general groups based on their functionality:

1. **Database Management Tools:**
   - **SQL Server Management Studio (SSMS):** SSMS is the primary tool for managing SQL Server databases. It provides a rich graphical interface for database administration, including query optimization, performance tuning, and database design.
2. **Performance Monitoring and Tuning Tools:**
   - **SQL Server Profiler:** SQL Server Profiler is a powerful tool for monitoring SQL Server activities, including query execution, stored procedures, and system events. It helps identify performance bottlenecks and troubleshoot issues.
   - **SQL Server Extended Events:** Extended Events is a flexible event-handling system for SQL Server that provides detailed monitoring and troubleshooting capabilities.
   - **SQL Server Query Store:** Query Store is a built-in feature that captures query execution plans and performance statistics over time. It helps identify and fix performance regressions and query plan issues.
   - **SQL Server Data Collector:** Data Collector is a built-in framework for collecting performance and system data, which can be used for monitoring and diagnosing issues.
3. **Query Analysis and Optimization Tools:**

➢**SQL Server Query Analyzer:** Query Analyzer is a component of SSMS that provides query tuning and optimization features. It includes tools for examining execution plans, indexing recommendations, and more.

➢**SQL Server Database Engine Tuning Advisor (DTA):** DTA is a tool that analyzes workloads and recommends performance-improving actions, such as creating or dropping indexes and partitioning tables.

➢**Plan Explorer:** Plan Explorer is a free tool by SentryOne that enhances query plan analysis and optimization. It provides a more detailed and user-friendly view of execution plans.

4. **Diagnostic Tools:**

➢**SQL Server Profiler and Extended Events:** In addition to monitoring, Profiler and Extended Events can be used for diagnosing issues by capturing events, errors, and performance metrics.

➢**SQL Server Error Logs and Event Viewer:** These built-in tools help track and diagnose errors, warnings, and critical events related to SQL Server.

5. **Backup and Recovery Tools:**

➢**SQL Server Management Studio:** SSMS includes features for backup and restore operations, helping in disaster recovery scenarios.

6. **High Availability and Disaster Recovery Tools:**

➢**SQL Server AlwaysOn Availability Groups:** This feature provides high availability and disaster recovery capabilities, with monitoring and management tools built into SSMS.

7. **Replication and Clustering Tools:**

➢**SQL Server Replication Monitor:** Replication Monitor is a component of SSMS that helps manage and monitor replication configurations.

➢**SQL Server Failover Cluster Manager:** This tool is used for managing SQL Server clusters for high availability.

8. **Third-Party Tools:** There are also many third-party performance monitoring and optimization tools available for SQL Server, such as SolarWinds Database Performance Analyzer, Redgate SQL Monitor, and Quest Spotlight on SQL Server.

POSTGRESQL PERFORMANCE TOOLS

PostgreSQL offers a variety of performance optimization, monitoring, and diagnostic tools to assist database administrators and developers in managing and tuning PostgreSQL database systems. These tools can be classified into several general groups based on their functionality:

1. **Database Management Tools:**

   ➢**pgAdmin:** pgAdmin is a popular open-source administration and management tool for PostgreSQL. It provides a user-friendly interface for database administration, including query optimization, performance tuning, and database design.

2. **Performance Monitoring and Tuning Tools:**

   ➢**pg_stat_statements:** This built-in module in PostgreSQL tracks and logs SQL query statistics, including execution times and query plans. It allows you to identify slow or resource-intensive queries that need optimization.

   ➢**pg_stat_activity:** PostgreSQL provides a view called `pg_stat_activity` that allows you to monitor active database connections, queries, and other session-related information.

   ➢**pg_stat_monitor:** An extension that provides real-time monitoring of various PostgreSQL metrics, including I/O, lock waits, and more.

   ➢**pgBadger:** While not a native tool, pgBadger is a popular open-source log analyzer that parses PostgreSQL log files to create detailed reports, helping you identify performance issues.

3. **Query Analysis and Optimization Tools:**

   ➢**EXPLAIN and ANALYZE:** PostgreSQL includes the `EXPLAIN` statement to examine query execution plans and the ANALYZE option to obtain actual execution statistics. This allows you to understand how PostgreSQL processes queries and optimize them accordingly.

   ➢**pg_hint_plan:** An extension that provides a hinting mechanism for PostgreSQL queries, allowing you to influence query planning without rewriting SQL statements.

   ➢**pgTAP:** While primarily a testing tool, pgTAP can be used to ensure that optimized queries still produce the expected results during performance tuning.

4. **Diagnostic Tools:**

➢ **pg_stat_statements:** Besides performance monitoring, the statistics collected by this module can also be used for diagnosing query performance issues.

➢ **pg_stat_activity:** You can use this view to diagnose issues related to blocked queries, long-running queries, or other session-related problems.

➢ **pg_log:** PostgreSQL logs contain valuable diagnostic information that can help identify and resolve database issues. Log analysis tools can be used to parse and analyze these logs.

5. **Replication and High Availability Tools:**

➢ **pgBouncer:** While not a native PostgreSQL tool, pgBouncer is a connection pooler that can improve database performance by managing and reusing database connections efficiently.

➢ **pgpool-II:** This is a middleware solution that offers connection pooling, load balancing, and high availability features for PostgreSQL.

6. **Third-Party Tools:** There are numerous third-party monitoring and optimization tools available for PostgreSQL, including Datadog, Zabbix, and SentryOne, which provide additional features and integrations for PostgreSQL monitoring.

MONGODB PERFORMANCE TOOLS

MongoDB provides a range of performance optimization, monitoring, and diagnostic tools to assist database administrators and developers in managing and tuning MongoDB database systems. These tools can be classified into several general groups based on their functionality:

1. **Database Management Tools:**

➢ **MongoDB Compass:** MongoDB Compass is an official GUI tool for MongoDB that provides a user-friendly interface for database administration, query optimization, and performance tuning. It includes a query profiler for analyzing query performance.

2. **Performance Monitoring and Tuning Tools:**

➢ **MongoDB Atlas:** MongoDB Atlas is a cloud-based database service that includes monitoring and optimization features. It offers real-time performance insights, automated scaling, and performance recommendations.

➢**mongotop and mongostat:** These are command-line tools included with MongoDB that provide real-time performance statistics, such as query execution times, connection information, and index usage.

➢**Database Profiler:** MongoDB includes a built-in database profiler that allows you to log and analyze database operations and queries, helping you identify slow or frequently executed queries.

➢**mongodump and mongorestore:** While primarily used for backup and restore, these tools can also help with performance optimization by allowing you to capture the state of the database for analysis and testing.

3. **Query Analysis and Optimization Tools:**

➢**db.collection.explain("executionStats"):** In MongoDB, you can use the `explain` method with the `"executionStats"` verbosity option to retrieve query execution details, including query plans and performance statistics. This helps you understand how MongoDB processes queries and how to optimize them.

➢**Indexes:** MongoDB provides indexing support to optimize query performance. Tools like the `db.collection.createIndex()` method help in creating and managing indexes for efficient data retrieval.

4. **Diagnostic Tools:**

➢**MongoDB Logs:** MongoDB logs contain diagnostic information that can be useful for troubleshooting and identifying performance issues. Analyzing these logs can help pinpoint problems in your MongoDB deployment.

➢**mongorepair:** This tool can be used for repairing a corrupted MongoDB database, which can be critical for maintaining database health and performance.

➢**Third-Party Diagnostic Tools:** Various third-party tools, such as MMS (MongoDB Monitoring Service) and Ops Manager, offer advanced diagnostic and monitoring features for MongoDB deployments.

5.**Replication and High Availability Tools:**

➢**MongoDB Replica Sets and Sharding:** MongoDB provides built-in features for replication and sharding to ensure high availability and scalability. Configuring and managing replica sets and sharded clusters is essential for performance and reliability.

6. **Third-Party Tools:** Numerous third-party monitoring and optimization tools are available for MongoDB, including Datadog, New Relic, SolarWinds, and Percona Monitoring and Management (PMM), which provide additional features and integrations for MongoDB monitoring.

MongoDB offers several query analysis and optimization tools to help you understand and improve the performance of your MongoDB queries. These tools provide insights into how queries are executed and suggest ways to optimize them. Here are the main query analysis and optimization tools in MongoDB:

1. **db.collection.explain("executionStats"):** The `explain` method is used to retrieve detailed execution information for a specific query. By appending `.explain("executionStats")` to your query, you can obtain a report that includes query execution time, the execution plan chosen by MongoDB, and various statistics related to the query. This tool is invaluable for understanding how MongoDB processes your query and where optimization is needed.

Example:

```
db.collection.find({ field: "value" }).explain("executionStats")
```

2. **Index Analysis:** Indexes are crucial for query performance in MongoDB. You can use tools like the `db.collection.getIndexes()` method to view the existing indexes on a collection and the `db.collection.createIndex()` method to create new indexes. Properly designed indexes can significantly improve query execution speed.

Example:

```
db.collection.getIndexes()
db.collection.createIndex({ field: 1 })
```

3. **Query Profiler:** MongoDB has a built-in query profiler that records slow or frequently executed queries. You can enable the profiler for a specific database and set a threshold for query execution time. Profiler logs can be reviewed to identify performance bottlenecks.

```
db.setProfilingLevel(1, { slowms: 100 })
```

Example (to enable the profiler for the current database):

4. **mongotop and mongostat:** These command-line tools provide real-time statistics about the state of your MongoDB deployment. `mongotop` shows

information about current read and write activity, while `mongostat` provides an overview of server statistics like connections, query times, and operations.

Example:

```
mongotop
mongostat
```

5. **Visual Query Analyzers:** Some third-party MongoDB management tools, like MongoDB Compass and Studio 3T, provide visual query analysis features. They allow you to profile and analyze queries using graphical interfaces, making it easier to understand and optimize query performance.

Example (using MongoDB Compass):

•Open MongoDB Compass.

•Connect to your MongoDB server.

•In the "Performance" tab, you can enable query profiling and analyze slow queries.

These tools collectively provide Database Administrators (DBAs) with the means to proactively monitor, diagnose, and optimize the performance of Oracle Database systems. Depending on the complexity of the environment and specific performance challenges, DBAs can choose the most suitable tools for their needs.

CHAPTER REFERENCES

(1) *IBM Documentation*. https://www.ibm.com/docs/en/db2-for-zos/12?topic=sql-statements (accessed 2023-09-26).

CHAPTER 9. SCALABILITY AND HIGH AVAILABILITY

Designing Scalable Database Architectures: Building for Growth and Performance

Designing scalable database architectures is indispensable for organizations looking to handle increasing data volumes, user loads, and evolving business requirements. A scalable database architecture allows for seamless growth, improved performance, and enhanced system responsiveness. This section highlights key principles and best practices for designing scalable database architectures that can effectively handle growing data demands and ensure optimal performance.

- **Data Partitioning and Sharding**: Partitioning and sharding techniques divide data across multiple physical or logical partitions or shards. This distributes the data and workload across multiple servers, enabling parallel processing and enhancing scalability. Partitioning can be based on various criteria such as range, list, or hash, depending on the data characteristics and access patterns.

- **Replication and High Availability**: Replication involves creating copies of the database on multiple servers to ensure high availability and fault tolerance. By replicating data, organizations can handle increased read loads, provide failover mechanisms, and improve system reliability. Techniques such as master-slave or multi-master replication enable data redundancy and allow for distributed processing.

- **Load Balancing**: Implement load balancing mechanisms to distribute incoming requests across multiple database servers. Load balancers evenly distribute the workload, preventing any single server from being overloaded. This ensures optimal resource utilization, improves response times, and enhances system scalability.[1]

- **Caching Strategies**: Caching mechanisms, such as in-memory caching or distributed caching frameworks, store frequently accessed data in memory. Caching reduces the need for repeated database access, improving response times and reducing the load on backend systems. By effectively leveraging caching, organizations can enhance scalability, handle high user loads, and improve overall performance.

- **Horizontal and Vertical Scaling**: Horizontal scaling involves adding more servers to a database cluster, distributing the workload across multiple nodes. Vertical scaling, on the other hand, involves increasing the resources (CPU, memory) of individual servers. Organizations should consider a combination of horizontal and vertical scaling approaches to meet their specific scalability requirements.

➢**Schema Design and Normalization**: Proper schema design and normalization help improve database performance and scalability. Normalize data to minimize redundancy and ensure data integrity. However, striking the right balance between normalization and denormalization is essential to avoid excessive joins and enable efficient data retrieval.

➢**Query Optimization and Indexing**: Optimize database queries to ensure efficient data retrieval. Analyze query execution plans, use appropriate indexing techniques, and fine-tune queries to minimize resource consumption. Well-designed indexes on frequently accessed columns can significantly improve query performance and overall system responsiveness.

➢**Continuous Monitoring and Performance Tuning**: Regularly monitor database performance, collect relevant metrics, and analyze system behavior. Proactively identify performance bottlenecks, fine-tune configurations, and optimize resource utilization based on real-time monitoring data. Continuous monitoring and performance tuning ensure that the database architecture evolves with changing demands.

Designing scalable database architectures is essential for organizations aiming to handle growing data volumes, user loads, and dynamic business requirements. By incorporating data partitioning and sharding, replication, load balancing, caching strategies, horizontal and vertical scaling, schema design, query optimization, and continuous monitoring, organizations can achieve scalable and high-performing database architectures. A well-designed scalable database architecture enables seamless growth, improved system responsiveness, and the ability to handle increasing data demands, setting the foundation for sustainable growth and success.

LOAD BALANCING AND REPLICATION STRATEGIES:

Ensuring Scalability and High Availability

Load balancing and replication strategies are fundamental components of designing scalable and resilient database architectures. Load balancing distributes incoming requests across multiple servers, ensuring optimal resource utilization and improved performance. Replication involves creating copies of the database on multiple servers, providing high availability, fault tolerance, and enabling distributed processing. This section emphasises the importance of load balancing and replication strategies in database design, highlighting key considerations and best practices for achieving scalability and high availability.

LOAD BALANCING:

➢**Even Workload Distribution**: Load balancing distributes incoming requests evenly across multiple database servers. By balancing the workload, organizations can avoid overloading individual servers, ensuring optimal utilization of resources and maintaining consistent performance even under high user loads.

➢**Horizontal Scaling**: Load balancing facilitates horizontal scaling by allowing organizations to add more servers to the database cluster. This enables the system to handle increased traffic and scale the workload across multiple nodes, providing improved scalability and responsiveness.

➢**Load Balancing Algorithms**: Different load balancing algorithms, such as round-robin, least connections, or weighted algorithms, determine how requests are distributed among the servers. The selection of an appropriate load balancing algorithm depends on factors such as server capacities, traffic patterns, and application requirements.

➢**Health Monitoring and Dynamic Load Balancing**: Load balancers typically include health monitoring mechanisms to check the availability and performance of database servers. In dynamic load balancing, servers that are experiencing high loads or exhibiting poor performance can be temporarily taken out of the rotation or have fewer requests routed to them, ensuring efficient resource utilization.

REPLICATION:

➢**High Availability and Fault Tolerance**: Replication creates multiple copies of the database on different servers, providing high availability and fault tolerance. If one server fails, another server can take over, ensuring uninterrupted service and minimizing downtime. Replication protects against data loss and ensures business continuity.

➢**Read Scalability**: With replication, organizations can leverage multiple database replicas to handle read-intensive workloads. By distributing read operations across replicas, organizations can scale the read capacity of the system, improving performance and response times for read-heavy applications.

➢**Replication Types**: Different replication types, such as master-slave or multi-master replication, offer varying degrees of flexibility and complexity. Master-slave replication involves one master server handling write operations, while multiple slave servers handle read operations. Multi-master replication allows for write operations on multiple servers, enabling distributed processing and better scalability.

➢**Data Consistency and Conflict Resolution**: Maintaining data consistency across replicated servers is vital. Organizations must implement mechanisms to ensure that updates made on one server are propagated to others in a consistent manner. Conflict resolution strategies, such as timestamp-based or conflict detection algorithms, are employed to address conflicts that may arise during replication.

Load balancing and replication strategies are key pillars in designing scalable and highly available database architectures. Load balancing evenly distributes requests across multiple servers, optimizing resource utilization and enhancing system performance. Replication ensures high availability, fault tolerance, and scalability, allowing organizations to handle increased workloads and provide consistent access to data. By implementing load balancing and replication strategies, organizations can achieve scalability, responsiveness, and resilience in their database systems. Considerations such as load balancing algorithms, health monitoring, replication types, data consistency, and conflict resolution techniques are crucial for successful implementation. Embracing these strategies empowers organizations to build robust database infrastructures that can meet the demands of growing data volumes, provide seamless user experiences, and ensure continuous availability of critical systems.

Clustering and Distributed Database Systems: Empowering Scalability and Data Availability

Clustering and distributed database systems are key architectural approaches that enable organizations to achieve scalability, fault tolerance, and high data availability. Clustering involves grouping multiple servers together to create a unified computing resource, while distributed database systems distribute data across multiple nodes to enable parallel processing and data redundancy. This section illuminates the importance of clustering and distributed database systems, highlighting their benefits, considerations, and best practices.

CLUSTERING:

1. **Scalability**: Clustering facilitates horizontal scaling by allowing organizations to add more servers to the cluster as the workload increases. This distributed approach enables efficient resource utilization, improved performance, and the ability to handle larger data volumes and user loads.
2. **High Availability and Fault Tolerance**: By clustering multiple servers, organizations achieve high availability and fault tolerance. If one server fails, others in the cluster can take over, ensuring uninterrupted service and minimal downtime. Clustering protects against hardware failures and provides a resilient infrastructure.
3. **Load Balancing**: Clustering often includes load balancing mechanisms that distribute incoming requests evenly across cluster nodes. Load balancers ensure optimal resource utilization, preventing any single server from being overloaded and maintaining consistent performance.
4. **Data Replication**: Clustering allows for data replication across multiple nodes, providing data redundancy and ensuring data availability. Replicating data across servers enhances fault tolerance, allows for local data access, and supports efficient data retrieval.

DISTRIBUTED DATABASE SYSTEMS:

- **Data Distribution and Partitioning**: Distributed database systems distribute data across multiple nodes, allowing for parallel processing and improved performance. Data can be partitioned based on various criteria, such as range, hash, or list, to achieve optimal data distribution and minimize data transfer during query execution.

- **Scalability and Performance**: Distributed database systems enable scalability by adding more nodes to the system. As data and workload grow, organizations can easily expand the cluster to accommodate increasing demands, ensuring efficient resource utilization and enhanced system performance.

- **Data Consistency and Synchronization**: Ensuring data consistency across distributed nodes is crucial. Distributed database systems employ synchronization mechanisms to propagate updates and maintain data consistency. Techniques such as two-phase commit (2PC) or distributed consensus algorithms help manage distributed transactions and maintain data integrity.

- **Latency Reduction and Data Locality**: Distributed database systems can be designed to minimize data transfer latency by locating data closer to the applications or users who access it. By considering data locality, organizations can reduce network latency and improve response times, particularly for geographically dispersed systems.

- **Data Security and Compliance**: Securing data in distributed database systems requires comprehensive security measures. Encryption, access controls, and data privacy mechanisms are requisite to protect data across distributed nodes. Compliance with relevant regulations and standards should also be considered when designing security measures.

Clustering and distributed database systems offer organizations the ability to achieve scalability, high availability, and efficient data management. Clustering allows for horizontal scaling, load balancing, and fault tolerance, ensuring optimal resource utilization and uninterrupted service. Distributed database systems distribute data across multiple nodes, enabling parallel processing, data redundancy, and improved performance. Considerations such as data distribution, scalability, data consistency, latency reduction, and data security are key to successful implementation. By leveraging the benefits of clustering and distributed database systems, organizations can build robust and scalable infrastructures that support their growing data needs, provide reliable access to critical data, and enhance overall system performance.

CLUSTERING AND DISTRIBUTED DATABASE SYSTEMS:

Empowering Scalability and Data Availability

Clustering and distributed database systems are key architectural approaches that enable organizations to achieve scalability, fault tolerance, and high data availability. Clustering involves grouping multiple servers together to create a unified computing resource, while distributed database systems distribute data across multiple nodes to enable parallel processing and data redundancy. This section emphasises the importance of clustering and distributed database systems, highlighting their benefits, considerations, and best practices.

CLUSTERING:

- **Scalability**: Clustering facilitates horizontal scaling by allowing organizations to add more servers to the cluster as the workload increases. This distributed approach enables efficient resource utilization, improved performance, and the ability to handle larger data volumes and user loads.

- **High Availability and Fault Tolerance**: By clustering multiple servers, organizations achieve high availability and fault tolerance. If one server fails, others in the cluster can take over, ensuring uninterrupted service and minimal downtime. Clustering protects against hardware failures and provides a resilient infrastructure.

- **Load Balancing**: Clustering often includes load balancing mechanisms that distribute incoming requests evenly across cluster nodes. Load balancers ensure optimal resource utilization, preventing any single server from being overloaded and maintaining consistent performance.

- **Data Replication**: Clustering allows for data replication across multiple nodes, providing data redundancy and ensuring data availability. Replicating data across servers enhances fault tolerance, allows for local data access, and supports efficient data retrieval.

DISTRIBUTED DATABASE SYSTEMS:

- **Data Distribution and Partitioning**: Distributed database systems distribute data across multiple nodes, allowing for parallel processing and improved performance. Data can be partitioned based on various criteria, such as range, hash, or list, to achieve optimal data distribution and minimize data transfer during query execution.

- **Scalability and Performance**: Distributed database systems enable scalability by adding more nodes to the system. As data and workload grow, organizations can easily expand the cluster to accommodate increasing demands, ensuring efficient resource utilization and enhanced system performance.

➢**Data Consistency and Synchronization**: Ensuring data consistency across distributed nodes is necessary. Distributed database systems employ synchronization mechanisms to propagate updates and maintain data consistency. Techniques such as two-phase commit (2PC) or distributed consensus algorithms help manage distributed transactions and maintain data integrity.

➢**Latency Reduction and Data Locality**: Distributed database systems can be designed to minimize data transfer latency by locating data closer to the applications or users who access it. By considering data locality, organizations can reduce network latency and improve response times, particularly for geographically dispersed systems.

➢**Data Security and Compliance**: Securing data in distributed database systems requires comprehensive security measures. Encryption, access controls, and data privacy mechanisms are key to protect data across distributed nodes. Compliance with relevant regulations and standards should also be considered when designing security measures.

Clustering and distributed database systems offer organizations the ability to achieve scalability, high availability, and efficient data management. Clustering allows for horizontal scaling, load balancing, and fault tolerance, ensuring optimal resource utilization and uninterrupted service. Distributed database systems distribute data across multiple nodes, enabling parallel processing, data redundancy, and improved performance. Considerations such as data distribution, scalability, data consistency, latency reduction, and data security are key to successful implementation. By leveraging the benefits of clustering and distributed database systems, organizations can build robust and scalable infrastructures that support their growing data needs, provide reliable access to critical data, and enhance overall system performance.

FAILOVER AND DISASTER RECOVERY PLANNING:

Ensuring Business Continuity and Data Resilience

Failover and disaster recovery planning are crucial components of any robust IT infrastructure. They provide organizations with the means to mitigate the impact of potential failures or disasters, ensuring business continuity, data availability, and minimizing downtime. This section features the importance of failover and disaster recovery planning, highlighting their benefits, considerations, and best practices.

FAILOVER PLANNING:

➢**Business Continuity**: Failover planning aims to ensure uninterrupted business operations in the event of a failure. It involves setting up redundant systems and infrastructure, such as redundant servers, networks, and data centers, to seamlessly take over in case of primary system failures. This redundancy guarantees continuous service provision and minimizes the impact of failures on business operations.

➢**High Availability and Resilience**: Failover planning enhances system availability and resilience by minimizing downtime. It involves establishing failover mechanisms, such as automatic failover or manual switchover, that swiftly redirect traffic or workload to backup systems. This ensures minimal disruption to services and maintains data availability for users.

➢**Data Replication and Synchronization**: Failover planning often includes data replication and synchronization mechanisms. By replicating data from primary systems to backup systems, organizations ensure that data is up to date and readily available in the event of a failover. Data replication can occur in real-time or near real-time, minimizing the risk of data loss.

DISASTER RECOVERY PLANNING:

➢**Data Protection and Recovery**: Disaster recovery planning focuses on safeguarding critical data and establishing procedures for data recovery in the face of major incidents, such as natural disasters, hardware failures, or cyberattacks. It involves implementing regular backups, offsite storage of backups, and establishing recovery procedures to restore data and systems to a functional state.

➢**RTO and RPO Considerations**: Recovery Time Objective (RTO) and Recovery Point Objective (RPO) are instrumental factors in disaster recovery planning. RTO defines the maximum tolerable downtime, while RPO defines the acceptable amount of data loss. Organizations must assess their requirements and establish backup and recovery strategies that align with these objectives.

➢**Testing and Validation**: Regular testing and validation of disaster recovery plans are essential. Organizations should conduct drills and simulated recovery scenarios to ensure the effectiveness of their plans. This includes validating data backups, testing recovery procedures, and training personnel on their roles and responsibilities during a disaster.

➢**Geographic Redundancy and Offsite Backups**: Disaster recovery planning often involves establishing geographically redundant data centers or utilizing cloud-based services for offsite backups. By storing data and infrastructure in different physical locations, organizations mitigate the risk of data loss or system unavailability caused by localized disasters.

Failover and disaster recovery planning are integral to maintaining business continuity and data resilience in the face of failures or disasters. Failover planning ensures high availability, seamless service provision, and minimal downtime by establishing redundant systems and failover mechanisms. Disaster recovery planning focuses on data protection, backup, and recovery strategies to restore systems in the event of major incidents. Considerations such as data replication, synchronization, RTO, RPO, testing, and geographic redundancy are crucial for effective planning. By implementing robust failover and disaster recovery plans, organizations can safeguard their critical data, minimize the impact of failures, and ensure the continuity of operations, ultimately protecting their business and maintaining the trust of their stakeholders.

CHAPTER REFERENCES

(1) K, S.; S, S.; P, Y.; K, G. Library Management System. *IJRASET* **2023**, *11* (8), 992–1011. https://doi.org/10.22214/ijraset.2023.55279.

CHAPTER 10.OPTIMIZING APPLICATION-DATABASE INTERACTION

EFFICIENT APPLICATION CODING PRACTICES

Efficient application coding practices are essential for optimizing the interaction between applications and databases. Well-designed and properly implemented code can significantly improve performance, reduce resource consumption, and enhance the overall efficiency of application operations. This section emphasizes key practices for efficient application coding that contribute to optimizing application-database interaction.

1. **Minimize Database Roundtrips**: Reduce the number of roundtrips between the application and the database by combining multiple database operations into a single transaction or query. Minimizing roundtrips helps reduce network latency and overhead, improving application performance.

2. **Use Prepared Statements and Parameterized Queries**: Utilize prepared statements and parameterized queries to prevent SQL injection attacks and enhance performance. Prepared statements allow for the reuse of pre-compiled SQL statements, reducing parsing and optimization overhead and improving query execution speed.

3. **Employ Caching Mechanisms**: Implement caching mechanisms to store frequently accessed data in memory. By caching data, the application can avoid unnecessary database queries, reducing the load on the database server and improving response times. However, ensure that cached data remains consistent and up to date.

4. **Efficient Data Retrieval**: Retrieve only the required data from the database instead of fetching unnecessary data. Select only the necessary columns and use pagination techniques to retrieve data in smaller chunks, especially for large result sets. This reduces the amount of data transferred and improves application performance.

5. **Optimize Data Manipulation Operations**: Batch database operations together to minimize the number of insert, update, or delete statements executed separately. This reduces the overhead of each operation and improves overall performance. Additionally, consider using bulk insert/update statements or database-specific mechanisms for optimized data manipulation.

6. **Proper Error Handling**: Implement error handling mechanisms for database-related exceptions. Gracefully handle errors and provide appropriate feedback to

users. Effective error handling ensures the application can recover from errors and provides valuable information for troubleshooting and debugging.

7. **Connection Pooling**: Utilize connection pooling techniques to manage database connections efficiently. Connection pooling allows the reuse of database connections, eliminating the overhead of establishing new connections for each database interaction. This reduces resource consumption and enhances scalability.

8. **Code Optimization**: Optimize application code for performance and efficiency. Avoid unnecessary loops, redundant calculations, or resource-intensive operations. Regularly review and refactor code to eliminate bottlenecks and improve overall application performance.

Efficient application coding practices play a vital role in optimizing the interaction between applications and databases. By employing practices such as minimizing database roundtrips, utilizing prepared statements, employing caching mechanisms, efficient data retrieval, optimizing data manipulation operations, proper error handling, connection pooling, and code optimization, organizations can improve application performance, reduce resource consumption, and enhance overall efficiency. These practices not only benefit the application's performance but also contribute to a more efficient use of database resources, ultimately providing a better user experience and improved scalability. By incorporating these coding practices, developers can optimize the application-database interaction, enabling applications to harness the full potential of the underlying database system.

ORM (OBJECT-RELATIONAL MAPPING) OPTIMIZATION:

Maximizing Efficiency in Application-Database Interaction

ORM (Object-Relational Mapping) is a widely adopted technique that bridges the gap between object-oriented programming and relational databases. It simplifies data access and manipulation by mapping database entities to corresponding objects in application code. While ORM provides convenience and abstraction, it can introduce performance overhead if not optimized properly. This section identifies key strategies and best practices for optimizing ORM usage, ensuring efficient application-database interaction.

1. **Careful Mapping and Model Design**: Efficient ORM optimization begins with careful mapping and model design. Establish a clear and concise mapping between application objects and database entities, considering relationships, associations, and data types. Avoid unnecessary mappings, minimize redundant data retrieval, and optimize relationships to avoid performance bottlenecks.

2. **Lazy Loading and Eager Loading**: ORM frameworks often provide lazy loading and eager loading options. Lazy loading defers loading related data until it is specifically requested, reducing initial query costs. Eager loading preloads

associated data, minimizing subsequent database trips. Properly choose between these strategies based on usage patterns and data access requirements to balance performance and resource consumption.

3. **Batch Loading and Prefetching**: ORM frameworks often support batch loading and prefetching mechanisms. Batch loading allows fetching multiple entities or collections in a single query, reducing database roundtrips. Prefetching fetches related data in advance, anticipating future data needs. Utilize these techniques to minimize database queries and improve data retrieval efficiency.

4. **Query Optimization**: Optimize ORM queries for efficient execution. Leverage ORM query features, such as query caching, query hints, and query optimization options. Utilize database-specific features like indexes, stored procedures, or database-specific query language extensions to enhance query performance. Analyze query execution plans and fine-tune queries to eliminate performance bottlenecks.

5. **Contextual Loading and Caching**: Apply contextual loading and caching strategies to minimize repeated database trips. Contextual loading involves loading related data within the same database context, avoiding additional queries. Caching frequently accessed data within the ORM framework or utilizing external caching mechanisms can significantly improve performance by reducing the need for repeated database access.

6. **Transaction Management**: Efficiently manage database transactions within the ORM framework. Use the appropriate transaction isolation level to balance data consistency and concurrency requirements. Minimize transaction scope and ensure proper commit and rollback handling to reduce resource contention and improve performance.

7. **Connection Pooling**: Leverage connection pooling techniques offered by the ORM framework to efficiently manage database connections. Connection pooling eliminates the overhead of establishing new database connections for each ORM operation, improving performance, and resource utilization.

8. **Profiling and Monitoring**: Regularly profile and monitor ORM performance to identify bottlenecks and areas for optimization. Utilize profiling tools and ORM-specific monitoring features to analyze database query performance, resource consumption, and identify potential areas for improvement. Continuous monitoring ensures ongoing optimization efforts and helps maintain high-performance ORM usage.

Optimizing ORM usage is fundamental for maximizing the efficiency of application-database interaction. By carefully mapping and designing models, utilizing lazy loading and eager loading techniques, optimizing queries, leveraging batch loading and prefetching, implementing contextual loading and caching strategies, managing transactions effectively, utilizing connection pooling, and conducting profiling and monitoring, organizations can enhance ORM performance and minimize resource consumption. Efficient ORM optimization enables applications to leverage the benefits of ORM while ensuring optimal data retrieval, improved scalability, and enhanced overall system performance. By implementing these optimization strategies, developers can unlock the full potential of ORM frameworks, achieving seamless application-database interaction and delivering high-performance applications.

Connection Pooling and Resource Management: Optimizing Database Connectivity

Connection pooling is a technique used to manage and reuse database connections, providing efficient and scalable connectivity between applications and databases. By maintaining a pool of pre-established connections, connection pooling minimizes the overhead of establishing new connections for each database interaction. This section emphasizes the importance of connection pooling and resource management in optimizing database connectivity, highlighting the benefits, considerations, and best practices.

1. **Efficient Resource Utilization**: Connection pooling improves resource utilization by reusing existing connections rather than creating new ones. Opening a new connection for each database operation can be resource-intensive and lead to unnecessary overhead. Connection pooling ensures that connections are utilized efficiently, reducing the strain on the database server and improving overall system performance.

2. **Connection Reusability and Scalability**: With connection pooling, connections are kept open and made available for reuse. Reusing connections eliminates the need to establish a new connection for each user request, significantly reducing the time and overhead associated with connection establishment. This approach improves scalability as it allows for the handling of a larger number of simultaneous user requests.

3. **Connection Management**: Connection pooling frameworks handle connection management tasks, such as connection establishment, validation, and timeout management. These frameworks monitor the health and availability of connections, ensuring that only valid and functional connections are provided to application components. This minimizes the risk of using broken or expired connections, enhancing application reliability.

4. **Connection Pool Sizing**: Determining the appropriate size of the connection pool is crucial. A pool that is too small can lead to connection bottlenecks and increased wait times, while an excessively large pool can consume excessive resources. Analyzing application requirements, anticipated user loads, and database server capacity helps determine an optimal pool size that balances resource utilization and performance.

5. **Connection Validation**: Connection pooling frameworks often include connection validation mechanisms to verify the health and validity of pooled connections. Regularly validating connections helps identify and remove any stale or broken connections from the pool. Proper connection validation ensures that only functional connections are utilized, reducing the likelihood of errors and improving overall system stability.

6. **Connection Timeout and Recycling**: Setting appropriate connection timeout values helps manage resource allocation and prevent resource depletion due to idle connections. Configuring idle connection timeouts allows connections that have been idle for an extended period to be closed and released back to the pool. Recycling idle connections ensures efficient resource utilization and prevents unnecessary resource consumption.

7. **Connection Pool Monitoring**: Monitoring connection pool usage and performance is essential for identifying potential issues and optimizing resource management. Monitoring tools and frameworks provide insights into connection utilization, pool usage patterns, and performance metrics. Analyzing these metrics helps fine-tune connection pool settings, adjust pool size, and ensure optimal resource allocation.

8. **Connection Pooling in Distributed Environments**: In distributed environments with multiple application instances, coordination and synchronization of connection pooling become vital. Proper synchronization mechanisms are required to prevent race conditions and ensure consistent connection pooling across multiple application instances. Distributed connection pooling frameworks and techniques help manage connection pooling in these complex scenarios.

Connection pooling and resource management are essential for optimizing database connectivity, improving system performance, and enhancing scalability. By efficiently managing database connections, connection pooling reduces resource overhead, minimizes connection establishment time, and improves application responsiveness. Proper connection validation, timeout management, and connection recycling ensure the use of healthy and valid connections, enhancing system reliability. Analyzing and monitoring connection pool usage allows for optimization of pool size, resource allocation, and performance tuning. Implementing connection pooling best practices in distributed environments further ensures consistency and coordination across multiple application instances. By adopting effective connection pooling and resource management strategies, organizations can achieve efficient database connectivity, enhanced application performance, and improved overall system efficiency.

CACHING STRATEGIES AND TECHNIQUES:

Enhancing Performance and Scalability

Caching is a powerful technique that improves system performance and scalability by storing frequently accessed data in a cache. Caching reduces the need for repeated expensive computations or database queries, resulting in faster response times and improved overall system efficiency. This section features the importance of caching, highlighting various caching strategies and techniques that organizations can employ to optimize performance and enhance scalability.

1. **Data Caching**: Data caching involves storing frequently accessed data in memory for fast retrieval. This strategy reduces the need for costly database queries or computations. By caching data, organizations can significantly improve response times and reduce the load on backend systems. Data caching is particularly useful for read-heavy applications and data that doesn't change frequently.

2. **Result Caching**: Result caching involves caching the results of computationally expensive operations or complex queries. Instead of recomputing the results for every request, the cached result is served. Result caching is beneficial for operations that involve substantial processing time or involve multiple data sources. It reduces computational overhead and improves application responsiveness.

3. **Page Caching**: Page caching is commonly used in web applications, where entire HTML pages are cached and served to subsequent users. This approach eliminates the need for dynamic generation of the same page for each user, significantly reducing the load on web servers and database servers. Page caching is suitable for static or semi-static content, improving scalability and response times.

4. **Fragment Caching**: Fragment caching involves caching specific parts or sections of a page that are dynamic or frequently changing. Rather than caching the entire page, only specific sections that are computationally expensive or database-intensive are cached. This strategy strikes a balance between dynamic content and caching, providing faster rendering of frequently accessed components.

5. **Database Query Result Caching**: Caching the results of database queries is a common technique to avoid repeated execution of the same query. Query result caching improves response times for subsequent requests, especially for queries that return consistent results over a short period. This strategy reduces the load on the database server and enhances overall application performance.

6. **Cache Invalidation**: To ensure data consistency, cache invalidation mechanisms are essential. When underlying data changes, cached entries need to be invalidated or updated to reflect the latest data. Time-based invalidation, event-based

invalidation, or explicit invalidation mechanisms can be employed based on the specific application requirements. Proper cache invalidation ensures that users always receive the most up-to-date data.

7. **Distributed Caching**: In distributed environments or applications with multiple servers, distributed caching allows data to be shared across nodes. Distributed caching ensures that cached data is available and consistent across different servers, enabling scalability and load balancing. It reduces the need for inter-server communication and minimizes the load on backend systems.

8. **Cache Eviction and Expiration**: Cache eviction policies define how long data remains in the cache before it is removed or refreshed. Time-based expiration or eviction policies ensure that cached data remains relevant and up to date. Eviction policies, such as LRU (Least Recently Used) or LFU (Least Frequently Used), determine which cache entries are removed when the cache reaches its capacity.

Caching strategies and techniques play a vital role in optimizing system performance and enhancing scalability. By effectively employing data caching, result caching, page caching, fragment caching, database query result caching, and distributed caching, organizations can significantly improve response times, reduce computational overhead, and enhance overall application efficiency. Proper cache invalidation mechanisms ensure data consistency and accuracy. Additionally, employing cache eviction and expiration policies ensures that cached data remains relevant and up to date. By leveraging caching strategies and techniques, organizations can achieve faster response times, reduced server load, improved scalability, and enhanced user experiences. Caching is a valuable tool in the optimization toolkit that enables efficient data retrieval and computational efficiency.

CHAPTER 11.PERFORMANCE TESTING AND BENCHMARKING

Designing Effective Performance Tests and Benchmarking Database Performance

Designing effective performance tests and benchmarking techniques are crucial for evaluating and optimizing the performance of a database system. Performance testing helps identify bottlenecks, measure system capabilities, and validate the effectiveness of performance optimizations. This section emphasizes the importance of designing effective performance tests and benchmarking methodologies, including key rules-of-thumb to differentiate between good and bad performance.

DESIGNING PERFORMANCE TESTS:

- **Define Clear Objectives**: Clearly define the objectives of the performance test. Determine what aspects of the database system you want to evaluate, such as query execution times, transaction throughput, or response times. Setting clear objectives allows you to focus on specific performance metrics and optimize accordingly.

- **Select Appropriate Workload**: Choose a workload that accurately represents the expected usage patterns and scenarios of the database system. Realistic workloads mimic actual user behavior and provide valuable insights into system performance under normal operational conditions. Consider factors such as read-to-write ratios, data volume, and concurrent user loads.

- **Scalability Testing**: Include scalability testing to assess how the database system performs under increasing workloads. Test scenarios with incremental user loads, data volumes, or transaction rates to identify performance limits and scalability bottlenecks. Scalability testing helps determine if the system can handle future growth and increasing demands.

- **Consider Edge Cases**: Include edge cases and stress testing scenarios to evaluate the system's performance under extreme conditions. Test for scenarios such as high concurrent user loads, complex queries, or large data sets. Edge case testing helps uncover potential weaknesses or limitations that may impact performance in exceptional circumstances.

BENCHMARKING DATABASE PERFORMANCE:

➢**Establish Baseline Performance**: Establish a baseline performance measurement by executing performance tests on the database system under normal operational conditions. This baseline serves as a reference point for future performance evaluations and allows you to measure improvements or regressions accurately.

➢**Consistent Environment**: Ensure that the benchmarking environment remains consistent across multiple performance tests. Factors such as hardware specifications, network configurations, and software versions should be standardized to eliminate variations that can skew performance results. Consistency in the benchmarking environment ensures accurate performance comparisons.

➢**Use Multiple Metrics**: Measure performance using multiple metrics to gain a comprehensive understanding of the system's performance characteristics. Consider metrics such as response time, throughput, latency, and resource utilization. Comprehensive performance metrics provide a holistic view of the system's behavior and help identify areas for optimization.

➢**Comparison against Acceptable Thresholds**: Establish acceptable performance thresholds based on business requirements and user expectations. Use these thresholds as guidelines to evaluate the performance results. Comparing performance metrics against these thresholds helps identify deviations from expected performance levels and highlights areas that require improvement.

KEY RULES-OF-THUMB FOR PERFORMANCE EVALUATION:

1. **Consistency**: Performance should be consistent across multiple runs of the same workload. Consistency ensures that users experience predictable performance and minimizes the occurrence of outliers or irregular behavior.

2. **Responsiveness**: The database system should respond promptly to user requests, providing quick and predictable response times. Long response times or delays can indicate performance bottlenecks or resource constraints.

3. **Scalability**: The system should demonstrate the ability to scale with increasing workloads. As user loads or data volumes grow, the system should maintain acceptable response times and throughput, without significant degradation.

4. **Resource Utilization**: Efficient resource utilization is crucial. The system should leverage hardware resources optimally, avoiding excessive CPU usage, memory

leaks, or disk I/O bottlenecks. Proper resource management ensures efficient performance and scalability.

5. **Stability**: A well-performing database system should exhibit stability under varying workloads. It should not exhibit unexpected crashes, lockups, or inconsistencies that impact data integrity or availability.

Designing effective performance tests and benchmarking methodologies is essential for evaluating and optimizing database performance. By defining clear objectives, selecting appropriate workloads, considering edge cases, establishing baselines, ensuring consistent environments, using multiple metrics, and comparing against acceptable thresholds, organizations can accurately measure performance, identify bottlenecks, and optimize the system accordingly. Key rules-of-thumb such as consistency, responsiveness, scalability, resource utilization, and stability help discriminate between good and bad performance, allowing organizations to make informed decisions to improve their database system's performance. Effective performance testing and benchmarking are essential for achieving optimal performance, enhancing user experiences, and ensuring the successful operation of database systems.

BENCHMARKING METHODOLOGIES AND TOOLS:

Assessing and Improving System Performance

Benchmarking is a fundamental process for assessing and improving system performance. It involves comparing the performance of a system or component against established standards or competitors to identify areas for improvement and optimize overall performance. This section identifies benchmarking methodologies and tools used to evaluate system performance, highlighting their importance and key considerations.

BENCHMARKING METHODOLOGIES:

1. **Standard Benchmarking**: Standard benchmarking involves comparing the performance of a system against established industry-standard benchmarks. These benchmarks provide a reference point for evaluating system performance, ensuring fair comparisons across different systems or components. Standard benchmarking methodologies, such as SPEC (Standard Performance Evaluation Corporation) benchmarks, provide standardized workloads and metrics for performance evaluation.

2. **Internal Benchmarking**: Internal benchmarking compares the performance of different components or versions within the same system. It helps identify performance differences between configurations, software versions, or hardware upgrades. Internal benchmarking allows organizations to assess the impact of changes on system performance and make informed decisions about optimizations.

3. **Competitive Benchmarking**: Competitive benchmarking compares the performance of a system against direct competitors or industry peers. It provides insights into how a system performs relative to similar offerings in the market. Competitive benchmarking helps organizations understand their competitive position, identify performance gaps, and drive improvements to gain a competitive edge.

4. **Application-Specific Benchmarking**: Application-specific benchmarking focuses on evaluating system performance under specific application workloads. It involves designing performance tests that mimic real-world usage patterns and scenarios. Application-specific benchmarking allows organizations to optimize performance for specific use cases and ensure the system meets application requirements.

BENCHMARKING TOOLS:

1. **Synthetic Benchmarking Tools**: Synthetic benchmarking tools generate artificial workloads to evaluate system performance. These tools create standardized tests that simulate user interactions, database queries, or computational workloads. Popular synthetic benchmarking tools include Apache JMeter, Geekbench, and sysbench.

2. **Real-World Workload Capture Tools**: Real-world workload capture tools analyze and capture actual system usage patterns and transform them into repeatable benchmarks. These tools record user interactions, database queries, network traffic, or system resource utilization. The captured workloads can then be replayed for performance evaluation. Examples of real-world workload capture tools include ApacheBench and LoadRunner.

3. **Profiling and Monitoring Tools**: Profiling and monitoring tools help measure and analyze system performance in real-time. They collect and analyze performance metrics, resource utilization, and system behavior. Profiling tools like Perf and VTune provide insights into code-level performance bottlenecks, while monitoring tools like Nagios and Zabbix offer comprehensive system-level performance monitoring.

4. **Database-Specific Benchmarking Tools**: Database-specific benchmarking tools assess database performance and scalability. These tools simulate database workloads and measure performance metrics like throughput, latency, and transaction rates. Popular database-specific benchmarking tools include HammerDB, TPCC-mysql, and pgbench for PostgreSQL.

CONSIDERATIONS FOR BENCHMARKING:

1. **Reproducibility and Consistency**: Benchmarking results should be reproducible to validate findings and track improvements over time. Consistency in benchmarking

methodologies, hardware configurations, software versions, and workload parameters is crucial for accurate performance comparisons.

2. **Realistic Workloads**: Benchmarking should use workloads that closely represent real-world scenarios to obtain meaningful performance insights. Workloads should consider factors like read-to-write ratios, concurrent user loads, and data volume to accurately evaluate system performance under typical usage conditions.

3. **Metrics Selection**: Selecting appropriate performance metrics is vital to assess system performance effectively. Key metrics may include response time, throughput, latency, resource utilization, and scalability. Choosing the right metrics ensures a comprehensive evaluation of system performance.

4. **Test Environment**: The benchmarking environment should closely resemble the production environment to reflect realistic performance conditions. Hardware specifications, network configurations, and software versions should align with the production environment to eliminate variations that could impact performance results.

Benchmarking methodologies and tools are essential for evaluating system performance, identifying areas for improvement, and optimizing overall system efficiency. By employing standard benchmarking, internal benchmarking, competitive benchmarking, and application-specific benchmarking, organizations can gain insights into system performance, drive optimizations, and gain a competitive edge. Using synthetic benchmarking tools, real-world workload capture tools, profiling and monitoring tools, and database-specific benchmarking tools helps accurately measure performance metrics and evaluate system behavior. Considerations for reproducibility, realistic workloads, metric selection, and test environment ensure meaningful performance comparisons. By leveraging benchmarking methodologies and utilizing appropriate tools, organizations can continuously assess, optimize, and enhance system performance, ultimately improving user experiences and achieving business objectives.

ANALYZING AND INTERPRETING PERFORMANCE TEST RESULTS:

Insights for Performance Optimization

Analyzing and interpreting performance test results is a decisive step in optimizing system performance. Performance tests provide valuable data about the behavior, scalability, and efficiency of a system under varying workloads. This section highlights the importance of analyzing and interpreting performance test results, highlighting key considerations and best practices for extracting actionable insights.

1. **Performance Metrics**: Performance test results provide various performance metrics, such as response time, throughput, latency, error rates, and resource utilization. Analyzing these metrics helps understand the system's behavior and identify potential bottlenecks. Each performance metric provides a unique perspective on system performance and can help pinpoint areas that require optimization.

2. **Comparisons Against Baseline**: Compare performance test results against established baselines or previous test results to identify improvements or regressions. A baseline provides a reference point for measuring the impact of optimizations. Significant deviations from the baseline can indicate performance issues or unexpected behavior that requires further investigation.

3. **Identifying Bottlenecks**: Identify performance bottlenecks by analyzing performance metrics and their correlations. Look for metrics that show abnormal behavior or constraints, such as high response times, increased error rates, or excessive resource utilization. Identifying bottlenecks helps prioritize optimization efforts and focus on areas that have the most significant impact on system performance.

4. **Root Cause Analysis**: Perform root cause analysis to understand the underlying reasons for performance issues. Investigate the system's behavior, examine logs, and analyze relevant metrics to identify the root causes of poor performance. Common root causes include inefficient algorithms, excessive database queries, network congestion, resource contention, or suboptimal configurations.

5. **Scalability Analysis**: Analyze performance test results under different workload levels to assess system scalability. Determine how system performance scales as user loads, data volumes, or transaction rates increase. Look for signs of diminishing returns or performance bottlenecks at higher workloads. Scalability analysis helps determine if the system can handle increased demands and if optimizations are necessary to improve scalability.

6. **Correlation Analysis**: Perform correlation analysis to identify relationships between different performance metrics. Identify metrics that have a significant impact on others, as improving these metrics can have a cascading effect on overall performance. For example, high CPU utilization may be correlated with increased response times, indicating a need for CPU optimization.

7. **Performance Tuning Opportunities**: Based on the analysis of performance test results, identify specific performance tuning opportunities. This could include optimizing database queries, improving algorithm efficiency, adjusting resource allocations, or fine-tuning configuration parameters. Prioritize optimization efforts based on the impact of identified bottlenecks and their relevance to system performance goals.

8. **Documentation and Reporting**: Document the findings, analysis, and recommendations resulting from performance test analysis. Prepare comprehensive reports and share them with stakeholders, development teams, and management. Clear and concise reporting helps communicate the performance insights and guides decision-making for further optimization efforts.

Analyzing and interpreting performance test results is crucial for identifying performance bottlenecks, understanding system behavior, and optimizing system performance. By examining performance metrics, comparing against baselines, identifying bottlenecks, performing root cause analysis, analyzing scalability, conducting correlation analysis, identifying performance tuning opportunities, and documenting findings, organizations can gain valuable insights to improve system performance and enhance user experiences. A systematic and comprehensive approach to performance test analysis empowers organizations to make informed decisions, prioritize optimization efforts, and drive continuous improvement. Leveraging performance test results enables organizations to optimize system performance, meet performance goals, and ensure a high-performing and efficient system.

CAPACITY PLANNING AND PERFORMANCE MODELING:

Optimizing System Performance and Scalability

Capacity planning and performance modeling are essential processes for organizations to ensure optimal system performance, scalability, and resource allocation. Capacity planning involves estimating future resource requirements based on expected workloads, while performance modeling allows for simulation and analysis of system behavior under different scenarios. This section features the importance of capacity planning and performance modeling, highlighting their benefits and key considerations in optimizing system performance and scalability.

1. **Understanding Workloads and Usage Patterns**: Capacity planning begins with a thorough understanding of workloads and usage patterns. Analyze historical data, user behavior, and anticipated growth to determine the expected demand on the system. Consider factors such as concurrent user loads, data volumes, transaction rates, and peak usage periods. Understanding workloads is crucial for accurately estimating resource requirements and planning for future growth.

2. **Resource Identification and Allocation**: Identify the system resources required to support the anticipated workloads. Consider CPU, memory, disk space, network bandwidth, and other relevant resources. Based on workload analysis, allocate resources appropriately to ensure optimal performance and avoid resource bottlenecks. Capacity planning helps organizations determine the right balance between resource availability and system performance.

3. **Scalability Analysis**: Performance modeling enables organizations to simulate system behavior under varying workloads and scale levels. By creating models that capture the characteristics of the system and its components, organizations can evaluate performance metrics, such as response time, throughput, and resource utilization, under different scenarios. Scalability analysis helps identify potential bottlenecks, determine system limits, and optimize resource allocation to ensure the system can handle increasing workloads.

4. **Sensitivity Analysis**: Perform sensitivity analysis within the performance models to evaluate the impact of changes in different variables. Modify workload patterns, resource allocations, or system configurations to understand their influence on system performance. Sensitivity analysis helps identify key factors that significantly affect system performance, enabling organizations to make informed decisions and prioritize optimization efforts.

5. **Forecasting and Future Planning**: Capacity planning provides a foundation for future planning by estimating resource requirements and identifying potential capacity constraints. Organizations can use capacity planning data to forecast growth, plan hardware or infrastructure upgrades, and ensure the system remains scalable and performant. By understanding future resource needs, organizations can proactively address scalability challenges and avoid performance degradation.

6. **Realistic Workload Simulation**: Performance modeling relies on realistic workload simulation to accurately assess system behavior. Use real-world workload patterns, user scenarios, or workload generators to mimic actual usage patterns. Realistic workload simulation ensures that performance models capture the complexity and dynamics of real-world scenarios, providing more accurate performance predictions.

7. **Consideration of Non-Functional Requirements**: Capacity planning and performance modeling should consider non-functional requirements, such as response time targets, service-level agreements (SLAs), and user experience expectations. Performance models should be built to meet these requirements, allowing organizations to validate that the system can handle the anticipated workload while meeting performance goals.

8. **Continuous Monitoring and Optimization**: Capacity planning and performance modeling are iterative processes that require continuous monitoring and optimization. As the system evolves and new workloads emerge, organizations should regularly reassess capacity plans, update performance models, and optimize resource allocation to ensure ongoing performance and scalability.

Capacity planning and performance modeling are vital for optimizing system performance and ensuring scalability. By understanding workloads, identifying resource requirements, analyzing system scalability, conducting sensitivity analysis, and forecasting future needs, organizations can proactively address performance challenges and allocate resources effectively. Realistic workload simulation and consideration of non-functional requirements further enhance the accuracy of performance models. Continuous monitoring and optimization ensure that performance remains optimal as the system evolves. By employing capacity planning and performance modeling practices, organizations can achieve a high-performing, scalable system that meets user expectations and supports business growth.

CHAPTER 12.ADVANCED TOPICS IN DATABASE OPTIMIZATION

Data Compression and Storage Optimization: Maximizing Efficiency and Performance

Data compression and storage optimization techniques play a crucial role in managing the ever-increasing volumes of data and improving storage efficiency. These approaches reduce data size, minimize storage requirements, and enhance data access and retrieval speeds. This section illuminates the importance of data compression and storage optimization, highlighting their benefits, key techniques, and considerations for maximizing efficiency and performance.

1. **Data Compression**: is the process of reducing the size of data to occupy less storage space. It offers several advantages, including reduced storage costs, faster data transfer, and improved system performance. Various compression algorithms, such as lossless and lossy compression, are employed based on the data type, application requirements, and the level of acceptable data loss.

2. **Lossless Compression**: These compression techniques compress data without any loss of information. They retain the original data after decompression. Lossless compression is ideal for scenarios where data integrity is critical, such as databases, text files, and system configurations. Popular lossless compression algorithms include ZIP, GZIP, and LZ77.

3. **Lossy Compression**: These methods reduce data size by permanently discarding certain details deemed less important to human perception. Lossy compression is suitable for multimedia content, such as images, audio, and video files, where minor data loss is acceptable. Common lossy compression algorithms include JPEG, MP3, and MPEG.

4. **Data Deduplication**: this identifies and eliminates duplicate data instances, storing only a single copy of each unique data block. Deduplication significantly reduces storage requirements by eliminating redundant data. It is particularly beneficial for backup and archival systems where large amounts of redundant data are stored. Deduplication techniques include content-aware and hash-based deduplication.

5. **Storage Tiering**: this involves categorizing data based on its importance or access frequency and placing it on different storage tiers. Frequently accessed or critical data is stored on high-performance, expensive storage devices, while less frequently accessed data is stored on lower-cost, slower storage media. Storage tiering optimizes resource utilization and enhances overall system performance.

6. **Compression Algorithms Selection**: Selecting the appropriate compression algorithm depends on the data characteristics, storage medium, and performance requirements. Consider factors such as compression ratio, compression and decompression speeds, and the impact on system resources. Balance between achieving optimal compression and maintaining acceptable performance to ensure efficient storage optimization.

7. **Compression Format and Metadata**: Consider the choice of compression format and metadata to facilitate efficient data access and retrieval. Opt for compression formats that allow random access or partial decompression of data, enabling quick retrieval of specific data elements. Metadata management, including efficient indexing and cataloging, enhances searchability and retrieval speed.

8. **Data Lifecycle Management**: Implement data lifecycle management practices to optimize storage efficiency. Classify data based on its age, relevance, and access patterns. Archive infrequently accessed data, delete expired data, and migrate active data to appropriate storage tiers. Efficient data lifecycle management ensures that data is stored in the most suitable and cost-effective manner.

Data compression and storage optimization are essential for managing the exponential growth of data and maximizing storage efficiency. By applying lossless and lossy compression techniques, data deduplication, storage tiering, and effective compression algorithm selection, organizations can significantly reduce storage costs, enhance data transfer speeds, and improve system performance. Efficient metadata management, data lifecycle management, and selecting appropriate compression formats further optimize data access and retrieval. Implementing these strategies allows organizations to effectively manage their data while maintaining data integrity and meeting performance requirements. Data compression and storage optimization techniques are valuable tools in the data management toolkit, enabling organizations to maximize storage efficiency, reduce costs, and improve overall system performance.

DATABASE COLUMNAR COMPRESSION:

Trade-offs and Benefits of Data Compression

Database columnar compression is a powerful technique used to reduce storage requirements and improve query performance in data management systems. By compressing data at the column level, database columnar compression offers numerous benefits, such as reduced disk space usage, improved I/O efficiency, and faster query processing. However, there are trade-offs to consider when implementing data compression. This section features database columnar compression, its advantages, trade-offs, and considerations for achieving an optimal balance between storage efficiency and performance.

DATABASE COLUMNAR COMPRESSION:

Database columnar compression is a technique that focuses on compressing data column-wise rather than compressing the entire row or table. Instead of storing data in a row-oriented format, where all columns of a single row are stored together, columnar compression stores data in a columnar format, where values from the same column are stored together. This columnar storage enables more effective data compression and delivers several benefits:

1. **Improved Compression Ratios**: Columnar compression algorithms exploit the similarities and repetitive patterns within a column to achieve higher compression ratios compared to row-oriented compression. Column values with similar characteristics, such as repeated values, tend to compress well, resulting in reduced storage requirements.

2. **Enhanced Query Performance**: Columnar compression improves query performance by reducing I/O operations and memory bandwidth requirements. With columnar storage, queries can selectively access only the required columns, leading to the retrieval of less data from storage and decreased disk I/O. This optimized data access results in faster query processing and improved overall system performance.

3. **Predicate Pushdown Optimization**: Columnar compression enables efficient predicate pushdown optimization. Query engines can evaluate predicates, such as filtering conditions, at the compressed data level before decompressing, reducing unnecessary data retrieval and improving query performance. This optimization minimizes CPU overhead and speeds up query execution.

TRADE-OFFS OF DATA COMPRESSION:

While data compression offers numerous benefits, there are trade-offs that organizations must consider when implementing compression techniques:

1. **CPU Overhead**: Data compression involves CPU-intensive operations, including compression and decompression. Compressing and decompressing data requires computational resources, which can impact CPU utilization and introduce additional processing overhead. Organizations should assess the trade-off between storage efficiency and the potential impact on CPU performance.

2. **Write Performance**: Compressed data requires additional processing during write operations, as data must be compressed before storage. This compression overhead can impact write performance, especially for systems with high write rates or real-time data updates. It is important to evaluate the write performance impact and ensure it aligns with the system's requirements.

3. **Compression and Decompression Overheads**: While columnar compression improves query performance, decompressing data incurs overhead during query execution. Decompression requires CPU resources, and the time spent on decompression can impact query response times. Balancing the compression ratio benefits with the decompression overhead is essential to achieve optimal query performance.

4. **Trade-off between Compression Ratio and Query Performance**: Selecting the right compression algorithm and parameters involves considering the trade-off between achieving higher compression ratios and maintaining acceptable query performance. More aggressive compression algorithms may result in higher storage efficiency but could introduce greater decompression overhead and impact query execution times. Striking a balance between compression ratios and query performance is crucial.

CONSIDERATIONS FOR AN OPTIMAL BALANCE:

To achieve an optimal balance between storage efficiency and query performance, several considerations are essential:

1. **Workload Analysis**: Analyze the data access patterns, query requirements, and performance goals of the system. Understand the read and write patterns, the importance of query response times, and the acceptable trade-offs between storage efficiency and CPU utilization.

2. **Compression Algorithm Selection**: Choose the compression algorithm based on the characteristics of the data and the system requirements. Assess the compression ratio, CPU overhead, and query performance impact of different algorithms. Experimentation and benchmarking can help determine the most suitable compression approach for a specific workload.

3. **Hybrid Compression Approaches**: Consider hybrid compression approaches that combine different compression techniques for different columns or data types. For example, highly repetitive columns may benefit from dictionary-based compression, while numeric or temporal columns may benefit from more specialized compression techniques. A tailored approach to compression can maximize storage efficiency and performance gains.

4. **Performance Monitoring and Tuning**: Regularly monitor system performance and collect performance metrics to assess the impact of compression on query performance. Fine-tune compression settings and parameters based on the observed performance characteristics to optimize the balance between storage efficiency and query execution times.

Database columnar compression offers significant benefits in terms of storage efficiency, improved query performance, and optimized data access. By compressing data at the column level, organizations can reduce storage requirements, enhance I/O efficiency, and achieve faster query processing. However, trade-offs exist, such as CPU overhead, write performance impact, and compression and decompression overheads. Organizations must carefully evaluate these trade-offs, considering workload characteristics, compression algorithm selection, hybrid compression approaches, and ongoing performance monitoring and tuning. By striking an optimal balance between storage efficiency and query performance, organizations can leverage database columnar compression to achieve efficient data storage, enhanced query processing, and overall system performance.

PARALLEL PROCESSING AND QUERY PARALLELIZATION

Database optimization through parallel processing techniques plays a indispensable role in improving system performance and query response times. Parallel processing and query parallelization techniques are powerful methods for achieving efficient database operations. By leveraging multiple processing resources and dividing query execution tasks, parallelization enables faster query processing and improved overall system performance. This section illuminates the importance and techniques of parallel processing and query parallelization in database optimization, highlighting their benefits, considerations, and best practices.

PARALLEL PROCESSING:

Parallel processing involves dividing a task into subtasks that can be executed simultaneously on multiple processing resources. In the context of database optimization, parallel processing enables the execution of database operations concurrently, utilizing multiple CPUs or processor cores. This approach offers several advantages:

- **Improved Query Performance**: Parallel processing significantly reduces query response times by dividing the workload across multiple processors. By executing parts of a query in parallel, the overall processing time is greatly reduced. Complex queries, data-intensive operations, and analytical workloads can benefit significantly from parallel processing, resulting in faster query execution and improved user experience.

- **Increased Throughput**: Parallel processing allows for higher throughput by executing multiple queries or operations simultaneously. As multiple tasks run in parallel, the system can handle a greater workload, accommodating more concurrent users and processing more queries within a given timeframe. This increased throughput enhances system scalability and performance.

- **Utilization of Resources**: Parallel processing maximizes resource utilization by leveraging multiple CPUs or processor cores. It efficiently utilizes available processing power, making better use of hardware resources and reducing idle time. This utilization optimization leads to improved system efficiency and cost-effectiveness.

QUERY PARALLELIZATION:

Query parallelization focuses on breaking down a query into smaller, independent tasks that can be executed concurrently. This technique enhances query performance and database optimization in the following ways:

- ➢**Partitioning and Distribution**: Query parallelization involves partitioning data and distributing query processing across multiple nodes or threads. By dividing the data into manageable portions, each node or thread can process a subset of the data simultaneously. This parallel execution significantly speeds up query processing.

- ➢**Task Coordination and Synchronization**: Query parallelization requires efficient coordination and synchronization mechanisms to ensure that subtasks are executed correctly and results are merged accurately. Synchronization points are crucial to combine partial results and produce the final query output. Effective coordination ensures the integrity and consistency of query results.

- ➢**Data Locality and Network Considerations**: When parallelizing queries across distributed systems, data locality and network considerations are important. Minimizing data transfer across the network by placing processing closer to the data source improves query performance. Ensuring efficient data distribution and minimizing network latency contribute to overall optimization.

BEST PRACTICES FOR PARALLEL PROCESSING AND QUERY PARALLELIZATION:

To effectively optimize database performance using parallel processing and query parallelization, consider the following best practices:

1. **Query Analysis and Optimization**: Analyze queries to identify computationally intensive or data-intensive operations that can benefit from parallel processing. Optimize query structures, indexes, and access paths to maximize parallelism and minimize dependencies among subtasks.

2. **Task Granularity**: Break down queries into appropriately sized subtasks to achieve a balance between parallel execution and coordination overhead. Fine-grained parallelism may lead to increased coordination overhead, while coarse-grained parallelism may underutilize resources. Experimentation and performance testing can help determine the optimal task granularity.

3. **Resource Management**: Implement efficient resource management techniques to allocate processing resources effectively. Load balancing and task scheduling strategies ensure that processing resources are evenly distributed and utilized optimally. Monitor resource utilization to identify bottlenecks and optimize resource allocations accordingly.

4. **Fault Tolerance and Recovery**: Design fault-tolerant mechanisms to handle failures during parallel processing. Implement checkpointing, task monitoring, and recovery mechanisms to ensure that failed tasks can be restarted or rerouted without compromising query results or system stability.

5. **Scalability Considerations**: Consider scalability when parallelizing queries. As the system and workload grow, ensure that parallel processing scales accordingly. Evaluate the scalability of parallel execution plans, data distribution mechanisms, and coordination strategies to avoid performance degradation as the system expands.

6. **Performance Monitoring and Tuning**: Regularly monitor query performance, system metrics, and resource utilization. Collect performance data and analyze bottlenecks or areas for improvement. Fine-tune parallelization settings, task distribution strategies, or query execution plans to optimize system performance continually.

Parallel processing and query parallelization are integral components of database optimization, enabling faster query execution, increased throughput, and efficient resource utilization. By harnessing multiple processing resources and dividing queries into smaller tasks, organizations can achieve significant improvements in system performance and user experience. Careful consideration of query analysis, task granularity, resource management, fault tolerance, scalability, and performance monitoring ensures effective parallelization and optimization. Leveraging the power of parallel processing and query parallelization empowers organizations to maximize the efficiency and performance of their database systems, ultimately enhancing productivity, scalability, and overall business success.

IN-MEMORY DATABASES AND CACHING FRAMEWORKS:

Boosting Performance and Scalability

In-memory databases and caching frameworks have emerged as powerful technologies for enhancing the performance and scalability of modern data-driven applications. By storing data in memory rather than on disk, these solutions offer significant advantages in terms of faster data access, reduced latency, and improved system responsiveness. This section stresses the importance of in-memory databases and caching frameworks, highlighting their benefits, use cases, and considerations for achieving optimal performance and scalability.

IN-MEMORY DATABASES:

In-memory databases store data entirely in the main memory, eliminating the need for disk I/O and leveraging the faster access speeds of RAM. They provide several key advantages:

➢**Speed and Reduced Latency**: By eliminating disk I/O, in-memory databases deliver exceptionally fast data access speeds. With data residing in memory, the time required to fetch and process information is significantly reduced, resulting in lower latency and improved application responsiveness.

➢**Improved Read and Write Performance**: In-memory databases excel in read and write operations due to the absence of disk-related latency. They enable high-speed data retrieval and seamless write operations, making them ideal for applications that require near-real-time data processing.

➢**Enhanced Scalability**: In-memory databases are well-suited for scaling horizontally, enabling seamless distribution of data across multiple nodes. This scalability allows applications to handle increasing workloads without sacrificing performance, making them an excellent choice for high-traffic or data-intensive systems.

CACHING FRAMEWORKS:

Caching frameworks, such as Memcached and Redis, complement in-memory databases by providing a caching layer between applications and databases. They offer the following benefits:

➢**Accelerated Data Access**: Caching frameworks store frequently accessed data in memory, allowing applications to retrieve data quickly without the need to query the underlying database. This caching mechanism dramatically reduces database access and improves application performance.

➢**Reduced Database Load**: Caching frameworks alleviate the load on databases by serving frequently requested data from memory. By reducing the number of database queries, they help improve database performance, optimize resource utilization, and scale the application more effectively.

➢**High Scalability and Availability**: Caching frameworks support distributed architectures, enabling seamless scaling and providing high availability. They can be deployed across multiple nodes or clusters, distributing the cache and ensuring that data remains accessible even in the event of failures or increased traffic.

CONSIDERATIONS FOR OPTIMAL PERFORMANCE AND SCALABILITY:

To maximize the benefits of in-memory databases and caching frameworks, consider the following considerations:

1. **Data Synchronization**: Ensure proper synchronization between the in-memory database and the underlying persistent storage to maintain data consistency. Implement mechanisms to update and synchronize data between the in-memory cache and the database, such as write-through or write-behind strategies.

2. **Cache Invalidation Strategies**: Implement efficient cache invalidation strategies to ensure that cached data remains accurate and up to date. Invalidating or refreshing the cache when data changes or expires helps maintain data integrity and prevent stale or outdated information from being served.

3. **Eviction Policies**: Define appropriate eviction policies within the caching framework to manage memory usage effectively. Consider factors such as data access patterns, data importance, and available memory. Eviction policies, such as least recently used (LRU) or least frequently used (LFU), ensure that the most relevant and frequently accessed data remains in the cache.

4. **Data Persistence and Durability**: Evaluate the need for data persistence and durability in the context of in-memory databases. Implement mechanisms, such as periodic backups or replication, to protect data against system failures or unexpected events. This ensures that critical data is not lost in the event of a power outage or system crash.

5. **Scalability and Distributed Architecture**: Design in-memory databases and caching frameworks with scalability in mind. Consider the distribution of data across multiple nodes or clusters to handle increasing workloads and accommodate future growth. Implement strategies such as sharding or partitioning to distribute data effectively and maintain high performance as the system scales.

6. **Monitoring and Performance Tuning**: Regularly monitor the performance of in-memory databases and caching frameworks to identify bottlenecks and optimize system performance. Collect and analyze relevant metrics, such as cache hit ratio, memory usage, and response times, to fine-tune configuration parameters and ensure optimal performance.

In-memory databases and caching frameworks have revolutionized the way data is stored and accessed, offering significant performance and scalability benefits for modern applications. By leveraging the speed and low latency of memory, in-memory databases deliver fast data access and improved read and write performance. Caching frameworks complement these databases by reducing database load and accelerating data retrieval. Considerations such as data synchronization, cache invalidation strategies, data persistence, scalability, and performance monitoring are essential for achieving optimal performance and scalability. The use of in-memory databases and caching frameworks empowers organizations to build high-performance, responsive, and scalable applications, ultimately enhancing user experiences and driving business success in the era of data-driven applications.

DATABASE OPTIMIZATION FOR BIG DATA PROCESSING:

Meeting the Challenges of Scale and Performance

As the volume, velocity, and variety of data continue to grow exponentially, database optimization for big data processing becomes increasingly crucial. Big data presents unique challenges, including massive datasets, complex processing requirements, and stringent performance demands. This section features the importance of database optimization for big data processing, highlighting key considerations, techniques, and best practices to achieve efficient and scalable data processing in this challenging environment.

CHALLENGES OF BIG DATA PROCESSING:

Big data processing poses several challenges that necessitate effective database optimization:

➢Data Volume and Velocity: Big data encompasses enormous volumes of data generated at high speeds. Traditional database systems may struggle to handle such large datasets and process data in real-time. Optimizations are required to efficiently manage and process these massive volumes of data without sacrificing performance.

➢Variety and Complexity: Big data encompasses diverse data types and structures, including structured, unstructured, and semi-structured data. Traditional relational database models may not be suitable for handling such diverse data types. Optimizations should account for the complexity of data models and enable flexible processing techniques to extract insights from a wide range of data sources.

➢Scalability and Distributed Processing: Big data processing often demands distributed computing environments to handle the scale of data and computational requirements. Optimizations must allow for horizontal scalability, effective distribution of data and processing across multiple nodes, and load balancing to ensure efficient utilization of resources.

KEY CONSIDERATIONS FOR BIG DATA DATABASE OPTIMIZATION:

To optimize database performance for big data processing, consider the following key considerations:

1. Data Partitioning and Sharding: Partitioning data across multiple nodes or shards improves data distribution and enables parallel processing. Choose appropriate partitioning strategies, such as range partitioning or hash partitioning, based on data characteristics and processing requirements. Efficient data partitioning enhances query performance, reduces data movement, and facilitates load balancing.

2. Distributed Query Processing: Leverage distributed query processing techniques to execute queries across multiple nodes in parallel. Optimizations, such as query decomposition, parallel execution plans, and data movement reduction, enable efficient distributed query processing. Distributing query processing across the cluster minimizes bottlenecks and improves overall performance.

3. Data Compression and Encoding: Apply data compression and encoding techniques to reduce storage requirements and improve I/O performance. Compressing data at rest or during transmission reduces disk space usage and network overhead. Encoding techniques, such as columnar storage or delta encoding, optimize data representation and enhance query performance.

4. Indexing Strategies: Choose indexing strategies that are suitable for big data processing. Traditional indexing techniques may not scale well for large datasets. Explore options such as bitmap indexes, compressed indexes, or distributed indexes that can handle high cardinality and large volumes of data more efficiently. Indexes should be carefully chosen to balance query performance and storage requirements.

5. Data Caching and Memory Management: Utilize in-memory caching frameworks to store frequently accessed data and improve query performance. Caching hot data in memory reduces disk I/O and enhances data retrieval speed. Efficient memory management, including buffer management and caching algorithms, optimizes memory utilization and reduces latency in data processing.

6. Stream Processing and Real-time Analytics: Incorporate stream processing frameworks to handle real-time data streams and perform continuous data analysis. Stream processing enables near real-time analytics, anomaly detection, and timely decision-making. Optimizations should focus on efficient stream processing, event time handling, and windowing techniques to extract insights from streaming data efficiently.

BEST PRACTICES FOR DATABASE OPTIMIZATION FOR BIG DATA:

To optimize database performance for big data processing, consider the following best practices:

1. Understand Data Patterns and Access Requirements: Thoroughly analyze data patterns, access requirements, and processing workflows. Identify data hotspots, access patterns, and specific processing needs to tailor optimizations to match these requirements effectively.

2. Perform Capacity Planning: Conduct capacity planning to estimate resource requirements based on anticipated data volumes, processing loads, and growth projections. Scale the infrastructure and database systems accordingly to accommodate future data growth and processing demands.

3. Embrace Parallel and Distributed Computing Technologies: Leverage parallel processing frameworks, distributed databases, and distributed file systems designed for big data processing, such as Hadoop, Apache Spark, or Apache Kafka. These technologies offer built-in optimizations for large-scale data processing and analysis.

4. Continuously Monitor and Fine-tune Performance: Regularly monitor database performance, collect performance metrics, and analyze system bottlenecks. Utilize monitoring tools to identify areas for improvement and fine-tune optimizations. Continuous performance tuning ensures that the system adapts to changing data characteristics and evolving processing requirements.

5. Emphasize Data Governance and Quality: Maintain data governance practices to ensure data quality, consistency, and reliability. Implement data cleaning, validation, and transformation processes to eliminate errors and inconsistencies that can impact processing performance and results accuracy.[1]

Database optimization plays a indispensable role in ensuring efficient and scalable big data processing. The challenges posed by massive data volumes, data variety, and high-speed processing requirements necessitate careful optimization strategies. By implementing data partitioning, distributed query processing, compression, indexing, caching, and memory management techniques, organizations can achieve optimal performance in big data processing environments. Embracing parallel and distributed computing technologies, continuously monitoring performance, and emphasizing data governance practices further enhance the efficiency and accuracy of data processing. Effective database optimization empowers organizations to unlock the value hidden within big data, enabling timely insights, data-driven decision-making, and gaining a competitive edge in today's data-intensive landscape.

CHAPTER REFERENCES

(1) Khakhar, D. Predictive Maintenance Strategies for Engineering Assets Using Data Analytics. *IJRASET* **2023**, *11* (7), 844–850. https://doi.org/10.22214/ijraset.2023.54747.

CHAPTER 13.FUTURE TRENDS IN DATABASE PERFORMANCE OPTIMIZATION

EMERGING TECHNOLOGIES AND THEIR IMPACT ON DATABASE PERFORMANCE

As technology continues to evolve at a rapid pace, emerging technologies are reshaping the database landscape and revolutionizing the way data is stored, processed, and accessed. These advancements introduce new possibilities for improving database performance, scalability, and efficiency. This section features several emerging technologies and their significant impact on database performance, highlighting their key benefits, challenges, and potential future directions.

1. **In-Memory Computing**: this is a game-changer for database performance. By storing data entirely in memory rather than on disk, in-memory databases eliminate the latency associated with disk I/O, resulting in significantly faster data access and processing. In-memory computing enables real-time analytics, faster transaction processing, and reduced response times. This technology has proven particularly beneficial for high-speed, data-intensive applications where low latency is necessary.

2. **Non-Volatile Memory (NVM)**: Non-Volatile Memory (NVM), such as NAND flash, 3D XPoint, or phase-change memory, is an emerging storage technology that combines the speed of memory with the persistence of storage. NVM provides high-speed access and low-latency data retrieval, blurring the line between memory and storage. Database systems leveraging NVM can achieve faster data writes and reads, reducing the I/O bottleneck and enhancing overall performance.

3. **Graph Databases**: Graph databases are designed to efficiently manage highly interconnected data, making them ideal for applications with complex relationships, such as social networks, recommendation systems, and fraud detection. Graph databases utilize graph-based data models and query languages optimized for traversing relationships. These databases excel at performing complex graph analytics and traversals, offering superior performance compared to traditional relational databases for graph-based workloads.

4. **Distributed Database Systems**: Distributed database systems leverage the power of distributed computing to process and store data across multiple nodes or clusters. They enable horizontal scalability, fault tolerance, and high availability. Technologies like Apache Cassandra, Apache Hadoop, and Google Spanner have revolutionized the handling of big data by distributing the workload and enabling

parallel processing, resulting in improved performance, scalability, and fault tolerance.

5. **Containerization and Microservices**: Containerization technologies like Docker and container orchestration platforms like Kubernetes are transforming the deployment and management of database systems. By encapsulating databases into lightweight, isolated containers, organizations can achieve efficient resource utilization, scalability, and agility. Microservices architecture, combined with containerization, allows database functionalities to be decoupled into smaller, manageable services, enabling better scalability, fault tolerance, and ease of deployment.

6. **Machine Learning and AI in Database Systems**: The integration of machine learning and artificial intelligence (AI) techniques within database systems opens up new possibilities for performance optimization. Intelligent query optimization and indexing, automated performance tuning, and predictive analytics capabilities enhance system efficiency and responsiveness. Machine learning algorithms can also be employed to detect anomalies, optimize resource allocation, and improve data management processes, ultimately leading to enhanced database performance.

CHALLENGES AND CONSIDERATIONS:

While emerging technologies offer numerous benefits, there are challenges and considerations to address:

1. **Adoption and Migration**: Integrating emerging technologies into existing database environments often requires careful planning and migration strategies. Organizations need to assess compatibility, data migration, and potential disruptions during the adoption process.

2. **Skill Set and Expertise**: Embracing new technologies may necessitate acquiring new skills and expertise. Organizations must invest in training and development to ensure the successful implementation and operation of emerging technologies.

3. **Data Security and Privacy**: As databases evolve, ensuring data security and privacy remains a foremost concern. Organizations must consider the implications of emerging technologies on data governance, compliance, and protection to maintain the integrity and confidentiality of sensitive information.

4. **Performance Tuning and Optimization**: While emerging technologies offer performance benefits, they may require specialized tuning and optimization approaches. Database administrators and developers must adapt their optimization strategies to harness the full potential of these technologies.

Emerging technologies are reshaping the database landscape and revolutionizing how organizations manage and leverage data. In-memory computing, NVM, graph databases, distributed systems, containerization, and AI integration are among the key advancements significantly impacting database performance. These technologies offer faster data access, improved scalability, fault tolerance, and advanced analytics capabilities. However, organizations must carefully consider the challenges associated with adoption, skill requirements, data security, and performance optimization. By embracing and effectively implementing these emerging technologies, organizations can unlock the full potential of their databases, achieving enhanced performance, scalability, and efficiency in the data-driven era.

CLOUD DATABASE PERFORMANCE CONSIDERATIONS:

Optimizing Data Management in the Cloud

The widespread adoption of cloud computing has transformed the way organizations manage and store their data. Cloud databases offer scalability, flexibility, and cost-efficiency, but optimizing performance in the cloud requires careful considerations. This section highlights the performance considerations specific to cloud databases, highlighting key factors, challenges, and best practices for achieving optimal performance in a cloud environment.

1. **Network Latency and Bandwidth**: One of the primary considerations for cloud database performance is network latency and bandwidth. Accessing data from a remote cloud server introduces additional network overhead, which can impact query response times. Optimizing network connectivity, selecting the appropriate cloud region, and utilizing content delivery networks (CDNs) can help mitigate latency and bandwidth limitations, enhancing performance.

2. **Scalability and Elasticity**: Cloud databases offer horizontal scalability, allowing organizations to scale their resources up or down based on demand. Leveraging auto-scaling features and distributed architectures, such as sharding or partitioning, enables the efficient utilization of cloud resources and ensures optimal performance during peak loads. Proactive capacity planning and continuous monitoring are crucial to maintain performance as data volumes and user demands fluctuate.

3. **Cloud Provider Selection**: Choosing the right cloud provider plays a significant role in database performance. Each provider offers different database services and infrastructure options, and performance characteristics may vary. Organizations should evaluate factors such as service-level agreements (SLAs), performance guarantees, infrastructure capabilities, and the provider's reputation for delivering consistent performance to make an informed decision.

4. **Data Placement and Replication**: Cloud providers offer various data placement options, including multi-region replication and availability zones. Placing data closer to the user or application can reduce latency and improve performance.

Organizations should strategically distribute data and consider replication options for high availability and disaster recovery while ensuring data consistency and minimizing synchronization delays.

5. **Database Configuration and Optimization**: Optimizing database configuration is crucial for performance in the cloud. Tuning parameters related to memory allocation, cache sizes, query optimization, and indexing can significantly impact performance. Regularly monitoring database performance metrics and fine-tuning configurations based on workload patterns and resource utilization ensures optimal performance in the cloud environment.

6. **Data Security and Compliance**: Cloud database performance should not compromise data security and compliance requirements. Organizations must ensure that proper security measures, such as encryption, access controls, and data privacy regulations, are implemented. Balancing performance optimizations with security considerations is vital to maintain data integrity and protect sensitive information.

7. **Hybrid and Multi-cloud Considerations**: these architectures offer flexibility and redundancy but require careful performance considerations. Balancing data distribution, network connectivity, and workload management across multiple cloud providers or a combination of on-premises and cloud environments is essential to maintain high performance. Optimizing data placement, connectivity, and data synchronization processes ensures efficient performance and seamless operations.

BEST PRACTICES FOR CLOUD DATABASE PERFORMANCE:

To optimize cloud database performance, organizations should follow these best practices:

1. **Regular Performance Monitoring**: Implement comprehensive performance monitoring to gain insights into database performance, resource utilization, and bottlenecks. Collecting and analyzing metrics enables proactive identification of performance issues and facilitates optimization efforts.

2. **Efficient Query Optimization**: Utilize query optimization techniques to improve query performance. Ensure proper indexing, query rewriting, and data modeling to optimize query execution plans. Implementing caching mechanisms or utilizing managed services that offer built-in query optimization capabilities can also enhance performance.

3. **Leveraging Managed Services**: Cloud providers offer managed database services that handle various administrative tasks, including performance optimization. Leveraging these services can offload maintenance responsibilities and allow organizations to focus on optimizing application performance and data access.

4. **Automated Scaling and Load Balancing**: Leverage automation capabilities to scale resources and balance workloads dynamically. Implement auto-scaling policies and load balancing mechanisms to handle varying traffic patterns effectively and optimize resource utilization.

5. **Continuous Performance Testing and Optimization**: Regularly test and optimize performance to identify bottlenecks and fine-tune configurations. Conduct performance testing under different workloads to validate database performance and identify areas for improvement.

6. **Regular Backup and Recovery Testing**: Implement robust backup and recovery strategies to ensure data availability and reliability. Regularly test backup and recovery processes to verify their effectiveness and minimize the impact of any potential data loss or system failures on performance.

A database configkey is a configuration key that controls the availability and functionality of a data entity or a data source in a database system. A data entity is a conceptual abstraction that represents a business object or a process, such as a customer, an order, or an invoice. A data source is a table, a view, or another data entity that provides data for a data entity.

A database configkey can be assigned to a data entity, a data source, or a field in either of them. Depending on the settings of the configkey, the data entity or the data source can be enabled or disabled, which affects the data management and integration operations that can be performed on them. For example, if the configkey for a data entity is disabled, the data entity will not be functional and cannot be used for import or export operations.

Some database systems use encryption keys to protect the data in the database from unauthorized access. Encryption keys are also configkeys that can be created, backed up, restored, deleted, or changed using the tools provided by the database system. Encryption keys can be applied to the whole database or to specific tables or columns within the database.

Optimizing performance in cloud databases requires a comprehensive understanding of the unique considerations and challenges of the cloud environment. Network latency, scalability, cloud provider selection, database configkey and data security are key factors to consider. By implementing best practices, such as regular monitoring, efficient query optimization, leveraging managed services, and embracing automation, organizations can achieve optimal performance in their cloud database deployments. With careful planning, continuous performance testing, and adherence to security and compliance requirements, organizations can unlock the full potential of cloud databases and deliver high-performance, scalable, and cost-effective data management solutions in the cloud.

AI AND MACHINE LEARNING FOR DATABASE OPTIMIZATION:

The emergence of artificial intelligence (AI) and machine learning (ML) has revolutionized various industries, and their impact on database optimization is no exception. AI and ML techniques offer new possibilities for improving database performance, query optimization, and resource management. This section features the role of AI and ML in database optimization, highlighting their key benefits, challenges, and potential applications for enhancing performance and efficiency.

1. **Query Optimization and Execution**: AI and ML algorithms can be leveraged to optimize query execution plans and enhance performance. By analyzing historical query patterns, query optimizer models can be trained to make intelligent decisions regarding join order, indexing, and data access paths. ML techniques, such as decision trees or reinforcement learning, can help in selecting the most efficient query execution plan based on data characteristics and workload patterns. This optimization reduces query response times and improves overall database performance.

2. **Automated Performance Tuning**: AI and ML can automate the process of performance tuning by continuously monitoring database metrics, workload patterns, and system resource utilization. Using anomaly detection algorithms, the system can identify performance bottlenecks and automatically adjust configuration parameters, such as buffer sizes, cache management, or indexing strategies, to optimize performance. This proactive and adaptive approach streamlines the performance tuning process and ensures efficient resource allocation.

3. **Predictive Analytics for Workload Management**: AI and ML techniques enable predictive analytics, allowing database systems to anticipate future workload demands and allocate resources accordingly. By analyzing historical data, system behavior, and workload patterns, ML models can forecast peak loads, data growth, or query spikes. This predictive capability facilitates proactive resource provisioning, capacity planning, and workload management, ensuring optimal performance during peak periods and avoiding resource shortages.

4. **Intelligent Data Indexing and Compression**: AI and ML algorithms can assist in intelligent data indexing and compression techniques. ML models can learn data access patterns and identify the most suitable indexes or compression algorithms to optimize storage and retrieval. This adaptive indexing and compression approach reduce disk space usage, improve I/O performance, and enhance data access efficiency.

5. **Anomaly Detection and System Health Monitoring**: AI and ML algorithms can detect anomalies in database performance and system health, enabling proactive

identification and resolution of issues. By monitoring various metrics, such as CPU usage, memory consumption, or query response times, ML models can learn normal behavior and identify deviations that indicate performance degradation or potential failures. Early detection allows administrators to take corrective actions, minimizing downtime and ensuring continuous optimal performance.

6. **Data Lifecycle Management and Data Placement**: AI and ML techniques aid in intelligent data lifecycle management, including data placement and migration strategies. ML models can analyze data access patterns, query patterns, and data characteristics to determine the optimal data placement and replication strategies. By placing frequently accessed data closer to the processing resources or considering data locality, AI-powered systems can reduce network latency and improve query performance.

CHALLENGES AND CONSIDERATIONS:

While AI and ML offer immense potential for database optimization, several challenges must be addressed:

1. **Data Quality and Bias**: AI and ML models heavily rely on quality data for training and decision-making. Ensuring data quality, integrity, and addressing biases in the training data is essential to maintain accurate and unbiased optimization outcomes.

2. **Model Interpretability**: Interpreting the decisions made by AI and ML models can be challenging. Understanding the reasoning behind query optimization or performance tuning decisions is crucial for administrators to trust and fine-tune the system's behavior.

3. **Resource Requirements and Scalability**: AI and ML techniques can be computationally intensive and require substantial computational resources. Organizations must consider the scalability and resource requirements of AI-powered database optimization solutions to ensure efficient performance in large-scale and high-traffic environments.

4. **Continuous Learning and Adaptation**: AI and ML models need to adapt to changing workloads, data patterns, and system configurations. Continuous learning and adaptation mechanisms must be implemented to keep the models up to date and ensure optimal performance as the database environment evolves.

AI and ML have the potential to revolutionize database optimization, offering intelligent techniques for query optimization, performance tuning, workload management, and resource allocation. By leveraging AI and ML algorithms, organizations can achieve faster query execution, proactive performance management, and efficient resource utilization. However, challenges such as data quality, interpretability, scalability, and continuous learning must be carefully addressed. As AI and ML techniques continue to advance, the integration of these technologies into database optimization processes will play a pivotal role in improving performance, scalability, and efficiency in the era of data-driven applications.

PREDICTIVE ANALYTICS DATABASE OPTIMIZATION:

Predictive analytics has emerged as a powerful tool in the field of database optimization and performance management. By leveraging historical data, statistical models, and machine learning algorithms, organizations can proactively optimize their databases, improve performance, and ensure efficient resource utilization. This section features the role of predictive analytics in database optimization and proactive performance management, highlighting its key benefits, challenges, and potential applications.

1. **Predictive Query Optimization**: Predictive analytics can be used to forecast query workloads and optimize query execution plans. By analyzing historical query patterns, workload variations, and data characteristics, predictive models can identify potential performance bottlenecks and suggest query optimization strategies. This proactive approach ensures that query execution plans are optimized in advance, reducing response times and improving overall database performance.

2. **Resource Allocation and Capacity Planning**: Predictive analytics enables organizations to forecast resource requirements and proactively allocate resources based on anticipated workloads. By analyzing historical data, workload patterns, and system behavior, predictive models can anticipate peak periods, data growth, or query spikes. This information enables administrators to provision resources in advance, ensuring optimal performance and avoiding resource shortages during peak demand.

3. **Anomaly Detection and Alerting**: Predictive analytics can detect anomalies in database performance and trigger proactive alerts. By continuously monitoring database metrics, such as query response times, CPU usage, or disk I/O, predictive models can learn normal system behavior. When deviations occur, such as sudden spikes in response times or abnormal resource consumption, the models can alert administrators, enabling them to investigate and address potential performance issues before they impact users.

4. **Proactive Performance Tuning**: Predictive analytics can help automate and optimize performance tuning processes. By analyzing historical performance data and identifying patterns, predictive models can recommend configuration optimizations, indexing strategies, or caching mechanisms. This proactive performance tuning ensures that database systems are continuously optimized for efficient resource utilization and improved performance.

5. **Workload Management and Query Prioritization**: Predictive analytics can aid in workload management by predicting query priorities and dynamically allocating resources based on criticality. By analyzing query patterns, business priorities, and service level agreements (SLAs), predictive models can recommend appropriate resource allocation strategies. This approach ensures that high-priority queries receive the necessary resources, reducing response times and meeting performance targets.

6. **Capacity Optimization and Data Archiving**: Predictive analytics can assist in capacity optimization by identifying unused or underutilized data. By analyzing data access patterns, data relevance, and historical usage, predictive models can recommend data archiving or tiering strategies. This optimization reduces storage costs, improves data retrieval times, and enhances overall system performance.

CHALLENGES AND CONSIDERATIONS:

While predictive analytics offers significant advantages in database optimization and proactive performance management, there are challenges to consider:

1. **Data Quality and Availability**: Predictive analytics relies on accurate and reliable data for modeling and forecasting. Ensuring data quality, integrity, and availability is essential for obtaining meaningful predictions and optimization recommendations.

2. **Model Training and Adaptation**: Predictive models need to be trained on relevant historical data and continuously updated to adapt to changing patterns and evolving database environments. Model training and adaptation processes must be implemented to ensure accurate predictions and proactive recommendations.

3. **Interpretability and Explainability**: Interpreting and explaining the decisions made by predictive models can be challenging. Organizations must ensure that models are transparent and provide insights into the factors influencing their recommendations to build trust among administrators and stakeholders.

4. **Integration and Scalability**: Integrating predictive analytics into existing database systems and workflows can be complex. Organizations must consider integration challenges, scalability requirements, and the need for real-time data processing to leverage predictive analytics effectively.

Predictive analytics has become a valuable asset in database optimization and proactive performance management. By leveraging historical data and advanced statistical models, organizations can anticipate workload variations, optimize query execution plans, and proactively allocate resources to ensure efficient performance. Despite challenges such as data quality, model training, and interpretability, the benefits of predictive analytics in database optimization are significant. By adopting a proactive approach and leveraging predictive analytics, organizations can enhance performance, improve resource utilization, and deliver a seamless user experience in their database environments.

PERFORMANCE OPTIMIZATION FOR GENERATIVE LARGE LANGUAGE MODELS

Key considerations for optimizing the performance of databases for large language models and context vectors are:

1. **Choosing the right vector database:** A vector database is a specialized type of database that is optimized for storing and performing operations on large amounts of vector data. Vector data are arrays of numbers that can represent complex data types such as text, images, voice, and structured data. Vector databases have advanced indexing and search algorithms that make them particularly efficient for similarity searches, a technique of searching for items most similar to a given item. This is one of the key requirements for augmenting prompts through contextual data in generative AI.[1]

2. **Indexing and partitioning the vector data:** Indexing and partitioning are techniques that improve the performance of vector databases by reducing the search space and the computational cost. Indexing involves creating a data structure that maps each vector to a unique identifier, which can be used to quickly locate the vector in the database. Partitioning involves dividing the vector data into smaller subsets, which can be distributed across multiple nodes or machines. This reduces the memory usage and the network latency, and enables parallel processing of queries.

3. **Tuning the query parameters:** Query parameters are settings that control how the vector database performs similarity searches. Some common query parameters are:

4. **Distance metric**: The distance metric is a function that measures how similar or dissimilar two vectors are. The most commonly used distance metrics are Euclidean distance, cosine similarity, and Jaccard similarity. The choice of distance metric depends on the type and distribution of the vector data, and the desired accuracy and speed of the query.

5. **Number of nearest neighbors**: The number of nearest neighbors (k) is a parameter that specifies how many similar items to return for a given query. The value of k depends on the use case and the trade-off between precision and recall. A higher value of k may increase recall, but also increase the query time and the noise in the results.

6. **Search algorithm**: The search algorithm is a method that determines how to find the nearest neighbors for a given query. There are various search algorithms, such

as hashing, quantization, or graph-based search, that use different techniques to optimize the search through approximation, compression, or traversal. The choice of search algorithm depends on the size and dimensionality of the vector data, and the desired accuracy and speed of the query.

CHAPTER REFERENCES

(1) MSV, Janakiram. How Large Language Models Fuel the Rise of Vector Databases. *The New Stack*. June 16, 2023. https://thenewstack.io/how-large-language-models-fuel-the-rise-of-vector-databases/.

ALPHABETICAL INDEX